DEC1R06

STATE OF THE SCIENCE: REVIEW OF QUANTITATIVE TOOLS TO DETERMINE WASTEWATER SOIL TREATMENT UNIT PERFORMANCE

by:

John McCray
Kathryn Lowe
Mengistu Geza
Jörg Drewes
Sarah Roberts
Assaf Wunsch
Colorado School of Mines

David Radcliffe
University of Georgia

Jose Amador
Janet Atoyan
Thomas Boving
University of Rhode Island

David Kalen
George Loomis
University of Rhode Island

2009

WERF

The Water Environment Research Foundation, a not-for-profit organization, funds and manages water quality research for its subscribers through a diverse public-private partnership between municipal utilities, corporations, academia, industry, and the federal government. WERF subscribers include municipal and regional water and wastewater utilities, industrial corporations, environmental engineering firms, and others that share a commitment to cost-effective water quality solutions. WERF is dedicated to advancing science and technology addressing water quality issues as they impact water resources, the atmosphere, the lands, and quality of life.

For more information, contact:
Water Environment Research Foundation
635 Slaters Lane, Suite 300
Alexandria, VA 22314-1177
Tel: (703) 684-2470
Fax: (703) 299-0742
www.werf.org
werf@werf.org

This report was co-published by the following organization. For non-subscriber sales information, contact:

IWA Publishing
Alliance House, 12 Caxton Street
London SW1H 0QS, United Kingdom
Tel: +44 (0) 20 7654 5500
Fax: +44 (0) 20 7654 5555
www.iwapublishing.com
publications@iwap.co.uk

Library of Congress Catalog Card Number: 2008942455

IWAP ISBN: 978-1-84339-358-0/1-84339-358-1

Colorado School of Mines

The research on which this report is based was developed, in part, by the United States Environmental Protection Agency (EPA) through Cooperative Agreement No. X-83085101-0 with the Water Environment Research Foundation (WERF). However, the views expressed in this document are solely those of [name of recipient] and neither EPA nor WERF endorses any products or commercial services mentioned in this publication. This report is a publication of WERF, not EPA. Funds awarded under the Cooperative Agreement cited above were not used for editorial services, reproduction, printing, or distribution.

ACKNOWLEDGMENTS

This research was funded by the Water Environment Research Foundation.

Research Team

Principal Investigator:
John McCray, Ph.D.
Colorado School of Mines

Co-Principal Investigators:
Jörg Drewes, Ph.D.
Mengistu Geza, Ph.D.
Kathryn Lowe
Colorado School of Mines

Project Team:
Sarah Roberts
Maria Tuchloke
Assaf Wunsch
Colorado School of Mines

David Radcliff, Ph.D.
University of Georgia

Jose Amador
Janet Atoyan
Thomas Boving, Ph.D.
David Kalen
George Loomis
University of Rhode Island

Issue Area Team

Steven Berkowitz
NC Department of Environmental & Natural Resources

Matt Byers, Ph.D.
Zoeller Company

Robert Freeman, P.E.
U.S. Environmental Protection Agency

Anish Jantrania, Ph.D., P.E.
NCS Wastewater Solutions

Charles McEntyre, P.E., CHMM
Tennessee Valley Authority

A. Robert Rubin, Ed.D.
McKim Creed

Edwin Swanson, P.E.
Arizona Department of Environmental Quality

George Tchobanoglous, Ph.D.
Tchobanoglous Consulting

Water Environment Research Foundation Staff

Director of Research: Daniel M. Woltering, Ph.D.
Senior Program Director: Amit Pramanik, Ph.D., BCEEM

ABSTRACT AND BENEFITS

Abstract:

A literature review was conducted as part of a larger research project to assess soil treatment unit (STU) performance in treating important wastewater constituents. Data were compiled from published field and laboratory studies relevant to fate and transport in soil of nitrogen (N), phosphorus (P), microbial pollutants, and emerging organic wastewater contaminants (OWC). The literature review also identified current best practices for using models and other tools to predict STU performance. The review demonstrates the variability of data collected at different field sites, and suggests that simple binary relationships (e.g., the ratio of a contaminant in soil to that in effluent, C/C_o, versus depth for various soil types) cannot accurately predict the attenuation of wastewater constituents in soil. In addition, many of the parameters that effect contaminant fate and transport in soils are not measured in available field studies. Most models for N and P fate and transport have been developed for agricultural applications, rather than for soil treatment of wastewater. However, the CW2D model contains many of the transformation processes relevant to STU performance and it may be possible to adapt the model for that use. Attempts to model microbial transport in soils have been hampered by the current incomplete understanding of the role of soil chemical and physical properties in transport and attenuation. Differences among the OWC, both in transport and treatment processes, suggest that no single model or modeling approach will be appropriate for all chemicals in this group.

Benefits:

- Compiles and evaluates over 200 data points from 84 experiments from the literature that describe N treatment and removal within STU.
- Highlights soil treatment processes that should be included in predictive tools to aid design of effective onsite wastewater treatment systems (OWTS).
- Describes factors that affect removal of various constituents in OWTS, including the presence of organics in soil water, soil-water saturation, and hydraulic loading rates.
- Presents evidence that different factors control the fate of different types of pathogenic organisms found in septic tank effluent, making it a challenge to optimize removal of these organisms in conventional soil-based systems.
- Reviews existing OWTS models and suggests priorities for development of an improved OWTS model.

Keywords: Soil treatment unit, onsite wastewater treatment, design, model, and performance

TABLE OF CONTENTS

LIST OF TABLES

LIST OF FIGURES

LIST OF ACRONYMS

Al	aluminum
ANIMO	Agricultural Nutrient Model
ANOVA	analysis of variance
BOD	biochemical oxygen demand
BOD_5	biochemical oxygen demand, five-day test
C	carbon
Ca	calcium
$CaCO_3$	calcium carbonate
CDE	convection dispersion equation
CEC	cation exchange capacity
CENTURY	Grassland and Agroecosystem Dynamics Model
CERES	Crop Environment Resource Synthesis
CO_2	carbon dioxide
CSM	Colorado School of Mines
CW2D	Constructed Wetlands 2D model
DNA	deoxyribonucleic acid
DNDC	Denitrification and Decomposition model
DO	dissolved oxygen
DOC	dissolved organic carbon
DOP	dissolved organic phosphorus
EPIC	Erosion-Productivity Impact Calculator
Fe	iron
FEM	finite element method
GC/MS	gas chromatography / mass spectrometry
HLR	hydraulic loading rate
IP	inorganic phosphorus
LAS	land application system
LDP	limited design parameter
LEACHM	Leaching estimation and Chemistry Model
MCL	maximum contaminant level
Mg	magnesium
N	nitrogen
N_2	dinitrogen gas
NH_4^+	ammonium ion
NLEAP	Nitrate Leaching and Economic Analysis Package
NO_2^-	nitrite
NO_3^-	nitrate
N_2O	nitrous oxide gas
NOE	Nitrous Oxide Emission
NRCS	Natural Resources Conservation Service
OH	hydroxyl
OM	organic matter
OWC	organic wastewater contaminants

OWTS	onsite wastewater treatment systems
P	phosphorus
pfu	plaque forming unit
PO_4^{3-}	phosphate
ppb	part per billion, microgram per liter
ppt	part per trillion, nanogram per liter
REMM	Riparian Ecosystem Management Model
RNA	ribonucleic acid
RNM	Riparian Nitrogen Model
RZWQM	Root Zone Water Quality Model
SDIS	subsurface drip irrigation system
SSURGO	Soil Survey Geographic
STE	septic tank effluent
STU	soil treatment unit
SWAT	Soil Water Assessment Tool
TMDL	total maxium daily load
TSS	total suspended solids
UGA	University of Georgia
URI	University of Rhode Island
US	United States
USDA	United States Department of Agriculture
U.S. EPA	United States Environmental Protection Agency
USGS	United States Geological Survey
VIRTUS	Virus Transport in Unsaturated Soils model
WARMF	Watershed Analysis Risk Management Framework

Executive Summary

Wastewater management in the United States relies on a variety of centralized and decentralized treatment approaches, with on-site treatment of wastewater often preferred for areas with lower density development. While onsite wastewater treatment systems (OWTS) vary widely in their design and implementation, conventional OWTS rely on the soil treatment unit (STU) for wastewater constituent treatment, hydraulic capacity, and eventual recharge to water resources. Regulatory guidelines for OWTS, however, often specify hydraulic performance with little consideration of treatment performance. Tools are needed to ensure that OWTS are designed, implemented, operated and maintained such that water quality is maintained and to prevent harmful levels of pathogens and trace organics from entering drinking water supplies.

The literature review described in this report is part of a larger research project to assess STU performance in treatming important wastewater constituents. The overall goal of the project is to provide a toolkit and tool-use protocol that is easy to implement and available to a wide range of users to assess STU performance. The review focused on conditions or factors that affect STU performance, and available models and other tools for predicting STU performance. Constituents of interest include nitrogen (N), phosphorus (P), microbial pollutants, and emerging organic wastewater contaminants (OWCs).

The authors reviewed the existing literature, including field and laboratory studies and modeling exercises, to assess current practice in predicting the performance of soil treatment units (STU) as a component of onsite wastewater treatment. Over 120 sources were reviewed, including articles from peer-reviewed journals, conference proceedings, and project reports. Of the sources reviewed, 25 studies containing 84 experiments were found suitable for N analyses and 11 studies describing 30 experiments were suitable for P analyses. Reported values were compiled for 16 parameters, including dimensions and age of the STU; hydraulic loading rate and loading regime; soil properties such as pH, organic matter content, and sand/silt/clay ratios; and effluent properties such as pH, BOD, COD and suspended solids.

The report summarizes the mechanisms that affect the fate and transport of nitrogen, phosphorus, and pathogenic microbes (bacteria, viruses, and protozoa) in soil. Available models are evaluated for their applicability to OWTS, including the degree to which they incorporate important transformation processes for wastewater constituents in soil. The report also examines the available data and models for predicting movement in soils of 5 priority organic contaminants that result from the use of pharmaceutical and personal care products.

The review demonstrates the variability of data collected at different field sites, and suggests that simple binary relationships (e.g., the ratio of a contaminant concentration in soil to that in effluent, C/C_o, versus depth for various soil types) cannot accurately predict the attenuation of wastewater constituents in soil. In addition, many of the parameters that effect contaminant fate and transport in soils are not measured in available field studies.

For nitrogen, hydraulic loading rate appears to be more important than soil texture or soil depth within the first 30-60 cm, although both soil depth and texture remain important variables. Given the variability of field data, multivariate statistical or empirical relationships likely will be needed to predict the performance of STU. However, in the frequent absence of field data, mathematical models are needed that incorporate relevant design variables and operating conditions.

Unlike nitrogen, which usually undergoes an irreversible process of mineralization-nitrification-denitrification in the STU, P participates in processes that can be reversible. Adsorption and precipitation of P are pH-dependent, with equilibrium affected by concentrations of P and other ions and the flow rate of wastewater in the STU. Most of the P-soil data are collected under equilibrium conditions, and do not investigate important kinetic behavior of phosphate compounds in soil. The few studies that examined P behavior in long-running STUs found that P plumes can form in groundwater under these systems because P is not retained in the soil treatment zone. Current data are not sufficient to determine P retention as a function of soil type. Greater understanding of P attenuation processes will be needed to improve modeling efforts.

Pathogen removal mechanisms vary because of the different physicochemical properties of the three groups: viruses, bacteria, and protozoa. Removal or inactivation of viruses occurs mainly by adsorption rather than mechanical filtration, due to their extremely small size. Thus, soil texture and depth are much less important for virus removal than soil pH and saturation. In contrast, removal of bacteria and protozoa appear to be most influenced by soil texture and depth. Optimizing removal of microbes in STU is challenging because of the different factors controlling the fate of the different types of microbes and because changes in these design parameters may alter the retention of nutrients such as N and P.

Organic wastewater contaminants, including those from pharmaceuticals and personal care products, are found in many wastewater streams but their fate and transport in soil environments has been little studied. Although it is known that organic molecules partition to organic carbon in the soil, field studies are needed to determine whether sorption is linear or nonlinear, an equilibrium or kinetic process, and whether or not the sorption is reversible. Bacterial degradation also can remove organic contaminants within an STU, but more research is needed to determine the importance of aerobic vs. anaerobic conditions for degradation, and the toxicity of degradation daughter products.

From the available studies for the 5 selected OWC—triclosan, 4-nonylphenol, 1,4-dichlorobenzene, 17β-estradiol, and diclofenac—the most effective STU design appears to be a shallow trench, which maximizes the soil surface area (and biomat formation) and enhances volatilization of OWC to air.

Few models have been developed to predict movement and treatment processes for N, P, or microbes in OWTS. However, the CW2D model, a module of the well-known HYDRUS model designed to simulate nitrogen treatment in a sand filter, contains most of the processes relevant to STU performance. CW2D includes a comprehensive treatment of microbial growth; the impact of oxygen mass transfer on nitrogen transformation; and variable rates of denitrification due to changes in dissolved oxygen concentrations, dissolved organic matter, and microbial growth. The review of existing models demonstrates that simulation of microbial characteristics in OWTS is largely uncharted territory.

Several studies have been conducted to model OWC fate and transport; however, it is unlikely that a single model or modeling approach will be appropriate for all OWCs because they form a broad class of compounds that undergo different transport and treatment processes.

CHAPTER 1.0

INTRODUCTION

1.1 Background and Motivation

Throughout the U.S., wastewater management incorporates a variety of centralized and decentralized approaches for protection of the public health and environment. Nearly 21% of the U.S. population is served by decentralized wastewater systems, with a substantial portion of all new development being supported by these systems (Lowe et al., 2007). As part of the overall wastewater management plan, proper onsite wastewater treatment system (OWTS) selection, design, installation, operation, and management are essential to ensure protection of the water quality and the public served by that water source.

While OWTS vary widely in their design and implementation, conventional OWTS rely on septic tanks for retention and digestion of gross solids in raw wastewater followed by discharge of wastewater effluent to the soil treatment unit (STU) for eventual recharge to underlying groundwater (Crites and Tchobanoglous, 1998; Metcalf and Eddy, 2003; Siegrist et al., 2001; U.S. EPA, 1997, 2002a). Figure 1-1 illustrates the processes that occur in OWTS. Engineered treatment units (e.g., sand filters, textile media filters) may be implemented where site conditions are not suitable for conventional systems or in sensitive areas such as those with nitrogen (N) loading concerns to produce a higher quality effluent delivered to the STU.

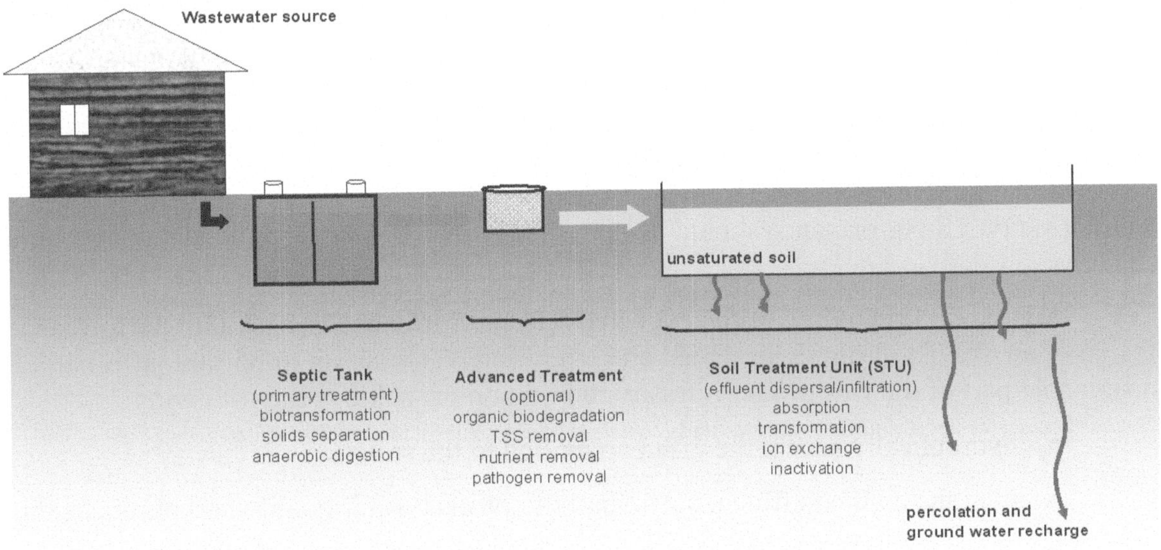

Figure 1-1. Schematic of a Conventional OWTS with STU.

In conventional systems, where local conditions permit, septic tank effluent (STE) (or higher quality effluent if additional treatment is employed) that may still contain high concentrations of pollutants, is further treated by discharging the effluent to the STU. A STU may be comprised of a series of subsurface trenches or beds for infiltration and percolation through an underlying unsaturated zone (vadose zone) with recharge to groundwater. An unsaturated flow regime may result in longer travel times and more extensive contact between percolating effluent and the soil. (In an unsaturated system, water is retained first in the finer pore spaces adjacent to soil grains and not in large pores.) In locations requiring additional soil depth (e.g., areas with only a thin layer of soil or a high water table), effluent may be discharged to an above-grade artificial mound. Unsaturated flow can be achieved within the STU by constraining design hydraulic loading rates (HLRs) to a fraction of the soil's saturated hydraulic conductivity. Effluent delivery to the STU can be achieved by gravity flow or dosed. At the soil infiltration surface, a biozone (also called a clogging zone and typically includes a biomat) can evolve and yield more uniform infiltration (both spatially and temporally) and enhance unsaturated flow (Siegrist and Boyle, 1987). This biozone, due to accumulation of microbial biomass and organic matter (OM), can enhance purification through sorption, biotransformation and die-off/inactivation processes.

Traditionally, OWTS design and regulation have been based primarily on ensuring that wastewater can be successfully infiltrated into the soil, preventing backup of the effluent to the soil surface or into the associated dwelling or business. However, this approach does not consider potential nutrient or pollutant mass loading to a receiving environment (soil, groundwater, surface water) in specific areas (single lot, subdivision, watershed). For example, in areas of growth, decisions are often made related to lower density suburban development served by OWTS compared to higher density urbanized development served by centralized treatment plants. In the past, prescriptive design and siting requirements were often based on the presumption that the system would perform to meet target goals. However, due to budget and staffing limitations, field systems are rarely monitored to verify that the STU is performing as expected with respect to treatment of wastewater constituents. Problems are typically highlighted only after a gross failure is observed (e.g., surfacing of effluent, detection of bacteria or nutrients in nearby drinking water wells or surface waters). Recently, decentralized cluster/ community systems have been recognized to offer higher density development options combined with the desired benefits of greenspace preservation, sustainable water resources, and lower infrastructure costs while maintaining high performance standards. In both low and high density development scenarios, OWTS performance should be specified in terms of achievable and measurable objectives.

A good understanding of the expected STU performance is critical to performance-based design of an OWTS. Properly designed and operating STUs are an important, affordable, and sustainable part of the wastewater infrastructure that must achieve the following:

- hydraulically accept the effluent applied to the soil;
- treat the applied effluent sufficiently to protect water quality and public health;
- provide a long service life with low operation and maintenance (O&M); and
- recover and reuse the water resource within the local watershed.

Understanding the treatment performance of soil in the STU for various OWTS system designs, and developing predictive tools to aid designers and decision makers, are critical for:

- assessing water quality impacts at the single site-scale and the watershed-scale;
- making informed management decisions to ensure protection of public health and the environment;

- achieving desired levels of treatment prior to recharge of the environment; and
- designing effective OWTS based on an understanding of STU processes.

The work presented here is part of a larger research project to assess STU performance with respect to treatment of important wastewater constituents: nitrogen (N), phosphorus (P), microbial pollutants (virus, bacteria, protozoa), and emerging organic wastewater contaminants (OWCs). The expected outcome is a toolkit and tool-use protocol that is easy to implement and available to a wide range of users. In some cases, more sophisticated tools (e.g., complex mathematical models) may be warranted depending on the relative complexity of the problem and the relative risk associated with a poorly designed STU. The tool-kit and protocol will focus on tools that are appropriate for practitioners, regulators, and planners engaged in STU design, but will also guide users toward more sophisticated modeling tools when necessary. All tools will be developed using rigorous experimental data and quantitative models verified with field data from operating systems.

1.2 Project Objectives

This report summarizes the available literature on best practices and available tools for STU design and performance evaluation for important wastewater constituents under a relevant range of OWTS operating conditions. The information and findings from this literature review will guide the next phase of the project, to develop STU design and performance monitoring tools for system designers and decision makers.

1.3 Project Approach

Available information on STU treatment performance was obtained from peer-reviewed journal publications, peer-reviewed conference proceedings (e.g., American Society of Agricultural Engineers [ASAE] now referred to as the American Society of Agricultural and Biological Engineers), and less widely distributed publications and project reports. Unpublished data were considered if provided by interested parties (e.g., the Florida Department of Health). The data were compiled and analyzed to determine key conditions affecting STU performance. The literature review also focused on current practices for using models and other available tools to predict expected STU performance.

Based on the information and insights from the literature review, field data from the three research sites (Colorado School of Mines [CSM], University of Rhode Island [URI], and University of Georgia [UGA]) will be utilized as needed to fill data gaps and/or develop the design tools to assess STU performance. The field monitoring framework and tool development will build from the base of existing knowledge presented here to evaluate the variations in STU performance under a range of design and soil conditions. The completion of this subsequent phase of the research project should provide new information on the performance of a given STU under various site conditions to remove key constituents from wastewaters with varying characteristics.

1.4 Report Organization

This report is organized into four chapters. The first chapter provides an introduction and purpose for the literature review. Chapter 2.0 describes STU performance based on data and information from the literature. The available modeling tools for constituent modeling within OWTS are presented in Chapter 3.0. Chapter 4.0 summarizes the analysis of the information reported in the literature and provides conclusions and recommendations for future monitoring.

CHAPTER 2.0

LITERATURE REVIEW OF SOIL TREATMENT UNIT PERFORMANCE

2.1 Introduction

Throughout the U.S., wastewater management incorporates a variety of centralized and decentralized approaches for protection of the public health and environment. Nearly 21% of the U.S. population is served by decentralized wastewater systems, with a substantial portion of all new development being supported by these systems (Lowe et al., 2007). Traditionally, OWTS design and regulation have been based on local practices, past experience, and soil percolation tests, despite known shortcomings. Design and siting requirements often presume that the system is performing to meet target goals. These regulations are often based on hydraulic performance with little or no rigorous consideration of treatment. In addition, research findings from geographic regions outside the local area or conducted in controlled laboratory environments are often discounted due to different environmental conditions or because general applicability of the findings is not readily understood or demonstrated.

While OWTS are typically robust treatment systems, problems can occur and have occurred due to site limitations and improper design. For example, there have been incidences of disease outbreaks by contaminated drinking water due to the source-water contamination (Craun, 1985; U.S. EPA, 2000). Some investigators have claimed that septic systems are the most frequently reported cause of groundwater contamination associated with disease in the U.S. (Powelson and Gerba, 1994). In addition, increasing attention has been focused in the past decade on the potential adverse effects to the environment and public health from the production, use and disposal of numerous synthetic and natural chemicals used in industry, agriculture, medical treatment, and common household conveniences (Daughton and Ternes, 1999), many of which are organic chemicals. Understanding the treatment performance of soil, its interaction with various OWTS system designs, and developing predictive tools to aid designers and decision makers, are critical for:

- assessing water quality impacts at the single site-scale and the watershed-scale;
- making informed management decisions to ensure protection of public health and the environment;
- achieving desired levels of treatment prior to recharge of the environment; and
- designing effective OWTS based on an understanding of STU processes.

This chapter will present the current state of knowledge on STU performance based on review and assimilation of reported data. The literature review focuses on nutrients (N and P), microorganisms, and OWCs from residential and commercial sources. Operational conditions and STU characteristics including HLR, soil type, effluent quality, study type (e.g., laboratory vs. field), existence and type of advanced treatment unit, and single versus clustered systems were captured to the extent that the data were available. Less information was available on microorganisms of interest, OWCs, and P compared to N. Microorganisms of interest included bacteria (fecal coliform, *E.coli*), viruses (indigenous coliphage), and protoza, although very very few data are available on protozoa. OWCs of interest include pharmaceutically active compounds, personal care products, and household chemicals. The findings from the literature review are presented by constituent, followed by discussions of experimental design and other factors that can affect the assessment of STU performance and the quality of the data.

2.2 Methods

To ensure results from the literature review were sound, available information was obtained from peer-reviewed journal publications, peer-reviewed conference proceedings (e.g., American Society of Agricultural Engineers [ASAE] now referred to as The American Society of Agricultural and Biological Engineers [ASABE]), and less widely distributed publications and project reports. Yet to be published data obtained from the project team were included where appropriate, but no attempts were made to collect unpublished data from individual researchers. The data were compiled and analyzed using various data-analysis techniques to help determine key conditions potentially affecting STU performance. Because the literature contains information reported in an array of formats for a wide range of conditions, the data were evalauted by individual constituent (e.g., N, P, microorganisms, and OWCs). A brief discussion of the methods used for data collection and handling is presented for each constituent in the following chapters. A similar concurrent literature review was conducted to assess the current best practices using models and other available tools to predict expected STU performance (presented in Chapter 3.0).

2.3 Nitrogen Treatment in Soil Treatment Units

2.3.1 Background

The main N species of concern in groundwater is nitrate (NO_3^-). Nitrate in drinking water has been associated with methemoglobinemia, which affects the ability of red blood cells to bind to oxygen (Shuval and Gruener, 1972). Infants are at greater risk when drinking NO_3^--rich water, hence the common name given to methemoglobinemia: the "Blue Baby Syndrome". In addition, NO_3^- concentrations above 4 mg N/L have been linked to Non-Hodgkin's Lymphoma (Ward et al., 1996). The United States Environmental Protection Agency (U.S. EPA, 1999) and the World Health Organization (WHO, 2006) have set a limit of 10 mg N/L in drinking water. Nitrate is also an important and limited nutrient in oceanic environments (Ryther and Dunstan, 1971). Discharge of NO_3^--rich groundwater to surface waters or the sea can lead to eutrophication – algal blooming – and deterioration of water quality (Brandes et al., 1974; Weiskel and Howes, 1992).

In the septic tank, organic N can be mineralized (or transformed) by microbes to ammonium (NH_4^+) (Figure 2-1). Ammonium can sorb to the soil or be utilized as an electron donor by nitrifying bacteria via the process of nitrification, when the STE is released to an aerobic environment. At the end of the nitrification process the NH_4^+ is converted to NO_3^-, with nitrite (NO_2^-) as a short-lived intermediate. Nitrate is a typically conservative anion which is readily leachable to groundwater (except at very low pH values where cation exchange of anions is possible). The NO_3^- in the soil pore water can be used as an electron acceptor in bacterial respiratory processes, and be converted to dinitrogen gas (N_2) or nitrous oxide gas (N_2O), thus resulting in net N loss from the system. This process, called denitrification, is dependent on several factors, such as saturation of the soil, availability of carbon source and suitable pH.

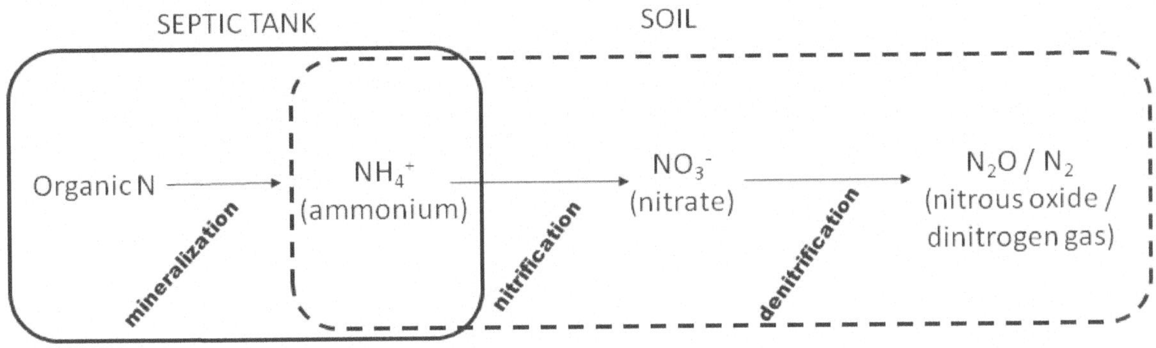

Figure 2-1. Main N Transformations in Septic Tanks and STUs (transformations are generalized, and may not be complete in field systems).

2.3.1.1 Nitrogen Transformations

Nitrogen as an element can be found in nature in a wide range of valence states, from the oxidized N^{+5} to the reduced N^{-3}. This enables the participation of nitrogen in a variety of compounds, organic and inorganic, and its utilization by microbes in several processes, both as an electron donor and acceptor. The main nitrogen processes include:

- Ammonification (or "mineralization") is the breaking down of organic nitrogen compounds, such as amino acids, to ammonia or ammonium with no valence change.
- Nitrification is the oxidation of inorganic nitrogenous compounds such as ammonium or ammonia. Nitrification is considered a two-stage process, each stage governed by a different group of bacterium. *Nitrosomonas* convert ammonium to nitrite, a short-lived intermediate nitrogen species, which in turn is converted to nitrate by *nitrobacter*. Both processes require oxygen as the electron acceptor. Ammonium or nitrite is the electron donor. Inorganic carbon is used by the microbes to build body mass.
- Denitrification is the reduction of oxidized nitrogenous compounds such as nitrate or nitrite to a gaseous phase. The denitrification chain is expressed as: $NO_3^- \rightarrow NO_2^- \rightarrow NO \rightarrow N_2O \rightarrow N_2$, although N_2 is not necessarily the end-product in all denitrification processes. The oxygen atom released in each stage is used to form CO_2. Nitrate is the electron acceptor, and organic carbon is the electron donor. Thus, sufficient organic carbon is required for this process. Denitrification is the primary mechanism for treatment of N in OWTS, so it is discussed in greater detail below.
- Nitrogen fixation is the result of some microorganisms that are capable of turning the usually-inert gaseous N_2 into ammonium or simple amino acids.
- Anammox, which stands for anoxic ammonia oxidation, is a process that was discovered in the mid-90s, and includes the oxidation of ammonium in an anaerobic environment, while using nitrate and nitrite as electron acceptors.

Of specific interest to STUs is nitrogen loss through denitrification. The denitrification process is carried out by heterotrophic bacteria. There are several requirements for the denitrification process to occur:

- Oxygen depletion is required because some denitrifiers are facultative (obligatory) anaerobes, and require an anoxic environment. Such an environment can be achieved when common heterotrophic bacteria, or nitrifying bacteria, consume the dissolved oxygen from the pore water, and the oxygen diffusion back into the water is limited. Oxygen diffusion to and dissolution into the pore water can be influenced by soil texture and water content. The process requires diffusion of gaseous oxygen through the soil air porosity, partitioning across the air-water interface, then diffusion in the aqueous phase to microbes. When the rate of diffusion to the water phase, and subsequent aqueous diffusion, is slower than the rate of consumption for aerobic bacteria, anaerobic conditions are acheived, and denitrification can occur. Higher water contents greatly reduce the rate of gaseous oxygen diffusion in the soil pore water, and also limit the air-water partitioning because there is less water-air interface to allow transfer of oxygen from the air into the water. Conditions beneficial to denitrification occur when the soil pores are at least 60% saturated, or when the soil air contains no more than 10% oxygen (these two factors are often related) (Brady and Weil, 2002). However, denitrification is believed to occur also in well-aerated soils, in anaerobic microsites (see McCray et al., 2005).
- Carbon source is necessary. When conditions are anoxic, nitrate is used as a final electron acceptor in the respiration process, instead of oxygen. In the denitrification process, organic matter is oxidized and carbon dioxide is released, much like in an oxygenic respiration process. Substantial denitrification may occur in "carbon hot-spots" in the soil, where decomposing roots or fauna act as carbon source (Parry et al., 2000). In OWTS, wastewater is the primary organic carbon source.
- pH for optimum denitrification is 7-8 (Martin and Focht, 1977).
- Temperature for optimum denitrification is 25-35°C, but the process will occur between 2 and 50°C (Brady and Weil, 2002).

2.3.1.2 Environmental Nitrogen and Plant Uptake

Nitrogen in the gaseous form of N_2 is the most abundant gas in the atmosphere, and comprises about 78% of all gases. In this form, nitrogen is inert and is not utilized by most living organisms, apart for specific groups of microorganisms that are able to fixate the inert gas and produce organic nitrogen compounds (although the chemical properties of N_2 make its utilization less common). Nitrous oxides, which may be produced in the process of denitrification, are highly reactive gases, which may contribute in a variety of ways to environmental pollution such as components of acid rain, green-house gases, or ozone destruction (Brady and Weil, 2002). Nitrogen losses to the atmosphere in the form of ammonia can be appreciable in certain applications. For example, algae in surface waters extract CO_2 from the water in the process of photosynthesis which drives up the pH of their aqueous environment. Under these conditions, ammonium ions in solution are naturally converted to ammonia, which volatilizes to the open air. STE is a minor source of ammonia volatilization, primarily due to subsurface discharge.

Nitrogen is important to all living forms since it is an essential component of amino acids, which are the building blocks of proteins. In plants, nitrogen is crucial for the assemblage of chlorophyll. A nitrogen-rich soil stimulates root growth and plant productivity, and the uptake of other nutrients. Overall, a healthy plant foliage can contain between 2.5% and 4% nitrogen (Brady and Weil, 2002). Plants might be useful for taking up nutrients from STUs, especially when the release of wastewater is shallow, such as in subsurface drip irrigation systems (SDIS). However, roots of vegetation also can clog and damage the STU (U.S. EPA, 2002b), disrupt biozones that form on the infiltrative surface, and cause preferential flow paths (e.g., rootholes) that short-circuit soil treatment

Potential sources of environmental nitrogen in soils are:

- Atmospheric deposition of nitrogen where nitrous oxides are released to the atmosphere through denitrification and fossil fuel combustion. These oxides are converted into nitric acid when dissolved in water during rainfall (Brady and Weil, 2002).
- Nitrogen fixation where conversion of dinitrogen gas to organic matter is completed by specialized soil bacteria (Lodwig et al., 2003).
- Wastewater application to STU through OWTS.
- Fertilization where artificial nitrogen fertilizers are applied to agricultural soils. Fertilizer made from this ammonia is estimated to be responsible for sustaining roughly 40% of the world's population and is the source for 40-60% of the nitrogen in the human body (Fryzuk, 2004).
- Geological sources where nitrogen is buried in rocks and released to the environment through erosion and volcanic activities. Although dinitrogen gas comprises 78% of the atmospheric gases, the atmospheric reservoir is only 2% of the nitrogen that exists on earth with the rest in geological sources (Brady and Weil, 2002).
- Feedlots where animal wastes release N to the soil surface.

2.3.1.3 Nitrogen Transport and Transformation Processes

Nitrogen transport occurs primarily via advective movement with the downward-flowing soil water. Other processes that influence the physical transport include dispersion and dilution in the subsurface. Transformations and reactions that occur along the flowpath are the primary means for "treatment". Thus, to understand the conditions that allow or interfere with nitrogen leaching from soils to groundwater, the affinity of the soil components to attract or repulse nitrogen ionic species must be understood.

Under neutral pH, most soil particles exhibit a negative net charge on the surfaces (Sposito, 1989) which allows the attraction of soil particles to positively charged ions, such as hydrogen protons (H^+) and ammonium. This adsorption is dependent on the charge and size of the ions in soil pore water, the charge of the soil particles and the available adsorption sites on the soil grains surface. The amount of adsorption sites depends on the grain size (generally, the smaller the soil particle, the more adsorption sites present) and the grain mineralogy. When the pH of the soil is reduced, hydrogen protons, which are positively charged, are attracted to the negatively charge surfaces of soil particles and reduce the ability of soil to adsorb more positively charged ions. On the other hand, at low pH, the soil-water interface is saturated with hydrogen protons, causing the soil to have a greater affinity to attract negatively charged ions.

Ammonium, a positively charged ion, has a good potential to adsorb to typically negative soil particles, and thus ammonium is less likely to leach into groundwater. Because of its small size, ammonium can become entrapped within cavities in the crystal structure of certain clays

(Brady and Weil, 2002). Soils with high amounts of clay minerals with a 2:1 alumina-silicate layering structure have a high capability of preventing ammonium mobilization in soil solution, as well as soils rich in humus (organic matter). In highly weathered (tropical) soils ammonium sorption is minor because 2:1 clay is usually not present. While adsorbed, the ammonium ions are held in exchangeable form, available for plant uptake (though slowly released), and partially protected from leaching. Apart from sorption, ammonium in the soil can undergo nitrification, assimilation by microbes for biomass building, or uptake by plant roots.

In contrast to ammonium, nitrate is highly mobile in soils under neutral pH due to its negative charge. Near-neutral pH is typical below OWTS because the pH of STE is near neutral, and STE also has a relatively high alkalinity to buffer the pH in soils. The nitrate ions are repulsed by most particle surfaces, which makes the nitrate leaching potential high. When nitrate reaches groundwater, concentrations are reduced primarily due to dilution, although denitrification in aquifers has been documented when regions of reducing geochemical conditions exist. In some hydrological systems, mostly in temperate regions with plenty of rainfall, groundwater is discharged to streams or lakes. The anoxic conditions that prevail in riparian zones and in river or lake sediments have a high denitrification potential (e.g., Mengis et al., 1999), along with plant uptake.

2.3.2 Data Collection and Handling Methods

Over 120 sources in the literature were reviewed (peer-reviewed journal articles, conference proceeding, project reports). In addition to the references cited in this report, a bibliography of additional works that were reviewed is presented in Appendix B. Of those sources, 93 references discussed nitrification or denitrification processes in soils. The main criterion for using a source in subsequent data analysis was the presence of reported values of N species concentrations vs. depth in STU. Of the literature sources reviewed, 25 sources containing 84 experiments were found suitable for data analyses. No attempt was made to solicite unpublished raw data beyond the project team, although raw data provided by interested sources (e.g., the Florida Department of Health) were evaluated.

Other parameters of interest that were compiled from the literature included: location of the experiment, field or laboratory classification, soil type, wastewater characteristics, wastewater source (e.g., single-residential home, office), and HLR. Engineered pretreatment units and cluster systems were also considered in this analysis, but data from these categories were sparse in the sources reviewed (two sources for engineered pretreatment units and none for cluster systems related to soil/vadose zone treatment treatment). A list of parameters extracted from the experiments, when available, is presented in Table 2-1. Table 2-2 summarizes in more detail the number of experiments reported in the literature that contained information for each of several properties or factors of interest for evaluating N removal in STUs. Each one of these parameters, and sometimes the combination of several of them, might represent the dominant effect on N attenuation in the soil. However, all parameters were not reported in each of the reviewed sources. To enable analysis of the current understanding of STU performance, the decision to include or exclude a source was based on reporting of the following information in the source:

- N vs. depth measurements;
- soil type (or the ability to infer soil type based on the experiment's location); and
- vadose zone properties.

Within these three criteria, various species of N were reported, some sources failed to specify soil type, and often little information was reported on vadose zone properties. Parameters such as wastewater source, loading regime, etc. were considered as "added value" to enable additional data assessment, but not as criteria for elimination of a source or experiment. Some of these parameters, if missing, were estimated when possible as described below.

Table 2-1. Experiment Parameters Considered in this Work.

General parameters	Soil properties	Effluent properties
Season in which the experiment took place	pH	pH
STU age	Sand / silt / clay ratios	Alkalinity
STU dimensions	Mineral content	BOD
HLR	OM content	COD
Loading regime (continuous, dosing)		TSS
Depth to groundwater		TOC

Based on the data obtained from the literature, statistical evaluations were planned to identify general trends and establish relationships. Tools envisioned included graphical tools such as summary tables and cumulative frequency distribution graphs and mathematical tools such as univariate correlations (e.g., N removal by soil type).

The sparsely populated categories in Table 2-2 show the difficulty of establishing simple or significant statistical relationships between some of the factors that might influence N attenuation and the observed concentrations with depth. For example, it would be very interesting to examine the effect of BOD on nitrogen removal. However, only 42 experiments (the ones that reported BOD in Table 2-2) could be used in the analysis. If a statistically significant result was observed from these 42 experiments and the effect of effluent HLR was also considered, the number of experiments would be further reduced to between 2 and 30 assuming that all of the experiments that reported BOD also reported HLR. To overcome the limitations of small data sets with multiple parameters, multivariate statistical methods (e.g., two-way analysis of variance (ANOVA), analysis of covariance, principal component analysis) are required.

In summary, while many parameters were identified during this review, reported studies often lacked sufficient detail on important parameters, perhaps because engineering aspects of OWTS often are proprietary. Therefore the statistical efforts reported herein focused on the parameters that are widely reported, such as HLR, soil texture, initial N concentration, and effluent loading regime (continuous vs. dosing).

Information was extracted from the relevant sources to a multi-sheet Excel file (Microsoft 2006), for later statistical analysis. All references were maintained using the EndnoteX software (Thomson Corporation 1988-2006). Statistical analyses were conducted using the Minitab software (Minitab Inc). Conversion of some of the data reported in the literature was sometimes required to enable comparison among the many different data sets. In addition, various species of N were often reported which required assumptions to estimate total N concentrations for comparison across studies. The following chapters describe each parameter in more detail and how it was handled for use in this report.

Table 2-2. Number of Literature Sources Reporting Specific Experimental Properties by Soil Type.

		Clay Loam, Silty Loam, Silt, and Sandy Clay	Loam, Sandy Clay Loam, and Sandy Loam	Clay, Silty Clay, and Silty Clay Loam	Sand, and Loamy Sand	Organic
Soil Properties	Soil Type	3	16	5	55	5
	Organic C	-	7	2	14	4
	Soil pH	-	8	2	18	4
	% Sand/Silt/Clay	-	8	2	19	-
	Minerology	-	-	2	8	2
	Slope	-	3	-	7	-
Effluent Properties	pH	2	9	2	12	4
	Alkalinity	2	2	-	12	-
	COD	2	5	-	23	4
	BOD_5	3	9	2	26	2
	TSS	2	2	-	13	2
	TOC	2	1	-	8	-
Effluent Source	Residential - Single	2	7	3	22	3
	Residential – Multi	-	2	-	19	-
	Campground	-	-	-	4	-
	School	1	-	-	-	-
	Synthetic Effluent	-	-	-	3	-
	Municipal Treatment	-	-	-	2	2
	Office	-	-	-	4	-
	Dairy Shed	-	6	2	-	-
Operational Properties	STU Dimensions (Field)	2	5	-	25	2
	STU Dimensions (Lab)	1	8	2	18	2
	HLR (Field)	2	8	3	30	3
	HLR (Lab)	1	8	2	20	2
	System Age	2	6	3	21	1
	Use of Pretreatment Unit	-	1	-	1	-
	Use of Recirculating Gravel Filter	-	-	-	1	-
	Continuous Effluent Delivery	-	2	2	17	1
	Dosed Effluent Delivery	2	4	1	20	2
Soil Saturation	Seasonally Saturated	-	3	2	8	1
	Seldom Saturated	-	1	-	-	1
	Constantly Saturated (> 10 months/year)	-	-	1	3	-
	Depth to GW	2	4	-	29	3
	Depth to GW not reported	1	12	5	26	2
Experimental Setting	Field	2	8	3	35	3
	Lab	1	8	2	20	2
Region (Field Studies)	North-East	-	1	-	3	-
	South	-	4	3	18	1
	West	-	2	-	-	2
	Mid-West	2	1	-	7	-
	Outside USA	-	-	-	7	-

2.3.2.1 Nitrogen Concentrations

To assess N treatment performance within the STU, both the concentration in the wastewater source (C_o) and the concentration (C) in the soil water at a particular depth in the STU must be expressed in similar units. The performance is represented by C/C_0. Nitrogen in wastewater appears in three main forms: organic N, NH_4^+ and NO_3^-. Nitrogen concentrations were reported different ways in different articles. Aqueous concentrations are often reported as one or more of the following: total N, NO_3^--N, NH_4^+-N, NO_3^- plus NH_4^+, and total Kjeldahl N value (the sum of NH_4^+ and organic N). Total N is the sum of NO_3^-, NO_2^-, NH_4^+ and organic N. In this report, the attenuation of total N from the soil was evaluated to enable a consistent statistical comparison between different experiments. To estimate the total N concentration when not all of the components were reported, several assumptions needed to be made:

- If the organic N concentration in the wastewater effluent was not reported, then total N concentration was estimated using the reported NH_4^+-N concentration. The NH_4^+-N concentration was divided by 0.85 to achieve a calculated total N concentration value. This was based on the assumption that NH_4^+ constitutes about 85% of a typical STE with negligible NO_3^- and NO_2^- concentrations (Kirkland, 2001).
- Some sources reported only NO_3^- and NH_4^+ aqueous concentrations in the soil. In cases that lacked any data on organic N, the total N concentration in pore water was calculated as the sum of the reported species. This has little error, because organic N consists of about 15% in the STE, and this value diminishes even further in the soil.
- Nitrite-N concentrations, if reported, were combined with NO_3^--N concentrations to calculate the NO_3^- concentration. This adjustment is based on the very low values of NO_2^- found in both STE and STU soil pore water owing to the rapid and nearly complete conversion of NO_2^- to NO_3^- in most environmental systems.
- Studies conducted on N attenuation and transformation in soils have shown that most NH_4^+ present in wastewater is oxidized quite readily, and that NO_3^- is the dominant N species in soil pore water within few tens of cm below the STE infiltrative surface (Walker et al., 1973b; Kristiansen, 1981; Cogger et al., 1988; Fischer, 1999). Therefore, when NO_3^- concentration was the only N-species reported in soil pore water, it was assumed that this value represented the total N concentration.

A preliminary analysis was required for the four sources that reported STU N concentration values as mg/kg of soil, rather than mg/L of soil pore water. A conversion from mg/kg of soil to mg/L of soil pore water is possible using the bulk density [kg/L] of the soil, the porosity and the water content (or saturation level) of the soil. However, sorbed organic N compounds and NH_4^+ to soil particles can lead to false estimated concentrations of N species in soil pore water that are calculated from total soil concentrations (results in aqueous concentrations that are higher than true values). In other words, to calculate aqueous concentrations from total soil water concentrations, the soil-water partitioning coefficient (K_d) must be known, but K_d values are rarely reported in these studies. K_d cannot be reliably estimated because it can vary highly among different soils (McCray et al., 2005). In addition, the majority of the articles did not report water content in soils. Thus, because the federal regulatory limit of 10 mg/L of NO_3^--N applies to the concentration in water and not in soil, it was decided not to use articles which reported only mg/kg soil values. Only four sources were eliminated by this condition and are not expected to have a significant impact on the results presented in this report. However, these data have been included in the Excel spreadsheet to enable future use as warranted.

In all of the reviewed sources, concentrations of different N species referred only to the concentrations of the N element, and not the N compound (e.g., reported as NH_4^+-N, NO_3^-- N). Concentrations reported as mol/L of N were converted to mg/L using the N molecular weight of 14 g/mol. For instance, a concentration of 3140 micro-mol of N per liter is equal to 43.96 mg/L, using the following conversion:

$$C_N = 3140\frac{\mu mol}{L} = 3.14x10^{-3}\frac{mol}{L} \rightarrow 3.14x10^{-3}\frac{mol}{L}\bullet 14\frac{g}{mol} = 0.0439\frac{g}{L} = 43.96\frac{mg}{L}$$

2.3.2.2 Experiment Location

The experiment location is very important because different field locations are expected to have different seasonal temperature, precipitation, wastewater composition, and even water use. Chemical, biological and enzymatic reactions all depend on temperature. Nitrification and denitrification are both biologically initiated chemical reactions, and therefore temperature is expected to influence these important processes. Temperature may also be important in other respects. For example, an experiment conducted in the cold Wisconsin winter found that even with freezing above-ground temperatures, the subsurface effluent discharge area did not freeze. This was probably due to heat release from microbial processes related to wastewater constituents, the warm effluent temperature, and the snow cover insulation (Bohrer and Converse, 2001). Location also influences climatic factors, which sometimes depend on temperature. Temperature has a strong influence on evapotranspiration, which can influence the subsurface flow regime and gas diffusion. Precipitation can be strongly location-dependent. Infiltrating rain can lead to saturation of the soil, and to creation of conditions which favor or inhibit nitrification or denitrification. For instance, higher saturations can limit the diffusion of oxygen to soil pore water, and greatly reduce nitrification rates, but also create anoxic conditions that are required for denitrification. Water content also influences volatilization of NH_4^+.

2.3.2.3 Field vs. Laboratory Classification

In the laboratory, the researcher can control almost every aspect of the experiment, and thus identify with more certainty the factors that lead to the observed experiment results. However, controlled laboratory settings are not equivalent to the more complex conditions of a field experiment, which are more representative of actual conditions for a particular location. Field results may not lead to conclusive determinations regarding the factors that influence the results, and the results may not readily transfer beyond the specific field location conditions.

2.3.2.4 Soil Type

Soils vary in physical properties, such as permeability, bulk density, field capacity, structure, and particle-size distribution (texture), and in chemical properties, such as acidity, mineral composition and organic content. Thus, the transport and transformation properties for many contaminants vary by soil texture. Acidity, much like temperature, plays a role in microorganisms' ability to execute certain biological processes, yet it also controls the availability of certain compounds to the organism. For instance, NH_4^+ transformation to ammonia – a volatile compound – is a strong function of pH, with relative concentrations of soluble ammonia increasing from 0.1% at pH 6, to 1% at pH 7, 10% at pH 8 and about 50% at pH 9 (Saggar et al., 2004). Studies of pH effect on nitrification rates suggest highest nitrification rates at pH values of around 7.5 (Painter and Loveless, 1981; Strauss et al., 2002; Wong-Chong and Loehr, 1975). Fortunately, this pH is similar to typical STE concentrations. High OM content in the soil can be important to N attenuation, because a carbon source is required by heterotrophic bacteria for the denitrification process (Brady and Weil, 2002). Organic matter content typically is greater closer to the ground surface, and where the infiltrating wastewater should also provide sufficient organic content for denitrification. However, as carbon in the STE is utilized, soil carbon at deeper depths may be required for continued denitrification. The particle-size distribution of the soil relates directly to the soil pore size, which in turn affects the soil saturation (Brady and Weil, 2002), tortuosity of the soil, and may also influence the chemical reactivity of the soil because particle-size distribution is related to surface area. Tortuosity and water content influence the gas- and aqueous-phase oxygen diffusion rates into and out of the soil water, which can be a limiting factor for nitrification (which requires oxygen) and denitrification (which requires anaerobic, or anoxic, conditions). With respect to denitrification, soil type may be an appropriate surrogate for a suite of important soil properties (Tucholke, 2007).

Some sources failed to specify soil type. If direct correspondence with the author did not provide sufficient information, the soil type was estimated or obtained from one of the following methods. While limitations are noted, it was assumed that these methods yield the appropriate soil type.

- The exact location of field experiments (e.g., student housing in a certain university) or location of soil collection for soil used in laboratory experiments was used to identify the soil type using the U.S. Department of Agriculture's (USDA) Web Soil Survey (http://websoilsurvey.nrcs.usda.gov/app/WebSoilSurvey.aspx). The Web Soil Survey is an interactive application based on the Soil Survey Geographic (SSURGO) database, which provides typical spatial scales of 1:15,840, 1:20,000, or 1:24,000 (Perchel, 2003). SSURGO was developed by the Natural Resources Conservation Service (NRCS) – a subdivision of the USDA.
- Where sand/silt/clay ratios were given, soil types were assigned using the soil texture triangle (http://www.uwsp.edu/geo/faculty/ritter/glossary/s_u/soil_texture_triangle).
- Using the reported soil phase (e.g. "Grandsdale loam – a coarse-silty over sandy or sandy-skeletal, mixed frigid Calciorthidic haploxeroll") and the general experiment location (county level) to retrieve the soil type from the USDA's Data Mart (http://soildatamart.nrcs.usda.gov). The Data Mart, like the Web Soil Survey, was developed by the NRCS and is based on the SSURGO database, yet does not contain an interactive map. Instead, it allows the user to generate PDF reports for specific soil types. Information provided in these reports includes soil physical and chemical properties, acreage and proportional extent of the soils, and other information.

The distribution of the 84 experiments to soil textures as classified by the USDA is presented in Figure 2-2. No experiments were conducted in silt, silt loam, silty clay and sandy clay textures. One experiment reported "silty sand" soil texture, yet this term is not part of the classic soil classification, but rather taken from the world of sedimentary geology (see Chatenet et al., 1996). Silty sand refers to sediment with sand as a major constituent, with 25-50% silt. With no other data available, silty sand was combined with sandy loam, which correlates to the sand-to-silt ratios described above.

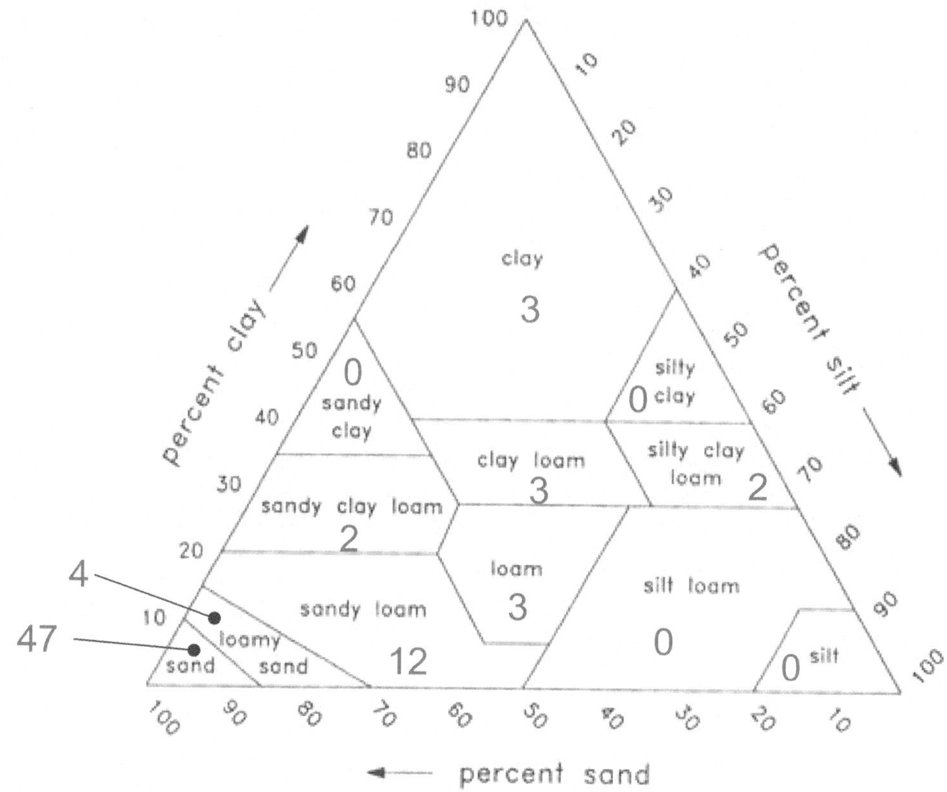

Figure 2-2. Soil Texture Triangle with Number of Experiments Conducted in Each Textural Type (red). A silty sand soil type, reported once in the literature review, was combined here with the sandy loam soil classification, due to lack of exact sand to silt ratios. The soil types reported as organic and gravelly sand are not included in this diagram.

Two other types of soils, "organic" and "gravelly sand", were reported. The term organic soil is not consistently defined in the literature, but rather there are several higher-level taxonomy terms which indicate a soil rich in organic matter (OM) (Brady and Weil, 2002). For example, "Melanic" refers to a soil that is thick, black, and high in OM (> 6% organic C), common in volcanic ash soils and "Histic" refers to a soil that has a very high organic content, is wet during some part of the year, and is usually formed in wet, boggy conditions. Five experiments conducted in soils with high organic content were reviewed in this work: four used the term "peat" and one simply referred to the soil as "organic". These experiments were grouped together into an "organic soil" category for the purposes of data analysis.

The gravelly sand descriptor is taken not from soil science, but from sedimentary geology (Folk's texture classification, see Blair and McPherson, 1999). This type of classification was encountered in two experiments: one from Perth (Western Australia), and one from Wisconsin. The Australian soil was defined as a "Chromosol" according to the Australian Soil Classification (CISRO website), which refers to soil with a "strong texture contrast" and pH >5. The addition of gravelly indicates that 10-20% of the soil volume consists of gravel (Australian Soil Collaborative Land Evaluation Program, CISRO). According to the article from which the experiment was taken, "A typical profile consists of an upper zone of ironstone gravels cemented together. This zone, termed 'caprock', can be up to 2 m thick, and is often overlain by loose sandy gravels ranging in depth from a few centimeters to more than a meter" (Gerritse et al., 1995). The Wisconsin soil was described as "single grain sand, with up to 60% gravel, cobbles and boulders" (Bohrer and Converse, 2001). Because of the uniqueness of these soil textures compared to the other types reviewed, they were kept as a separate category.

Figure 2-3 presents a distribution tree of the 84 experiments by soil type, field/laboratory setting and the loading regime of the effluent (continuous or dosing) for sandy and sandy loam soils. Figure 2-3 shows that every group of experiments with a common trait, such as soil type, can be subdivided based on other traits, which require special attention in statistical analysis. In addition to the distributions shown in Figure 2-2, the experiments can be divided further based on the criteria presented in Table 2-2.

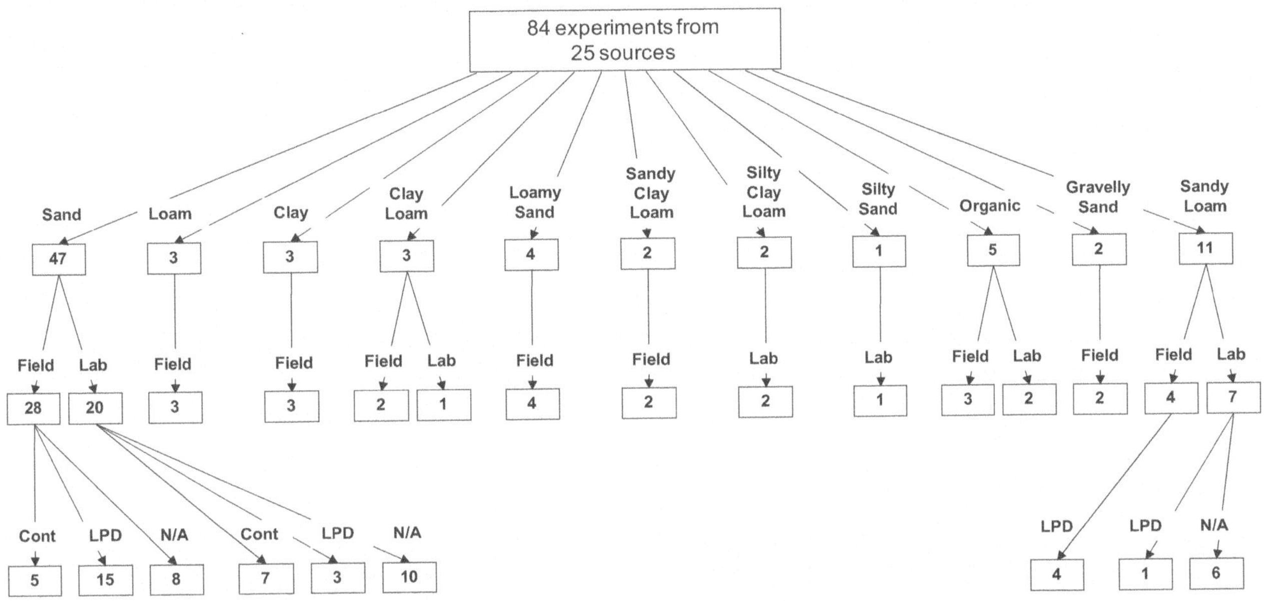

Figure 2-3. Distribution Tree of Soil Texture Types and Some Classifications Found in the Literature. Lab = laboratory experiments, Field = field experiments, Cont = continuous flow of septic effluent applied to soil, LPD = low-pressure dosing of effluent applied to soil, N/A = no information provided regarding discharge timing

Out of the 84 summarized experiments, 51 were conducted at only one depth. The single nitrogen concentration reported for one depth was reported as an average or median value of measurements taken over time. An additional 12 experiments reported nitrogen concentrations for two depths, and the remaining 21 experiments contain data from more than two depths.

2.3.2.5 Wastewater Characteristics

The most common species of N in STE is NH_4^+ (Crites and Tchobanoglous, 1998, U.S. EPA, 2004), mostly resulting from anaerobic decomposition of organic N compounds in the septic tank. In some systems (Bohrer and Converse, 2001) the STE goes through another treatment stage before being discharged to the soil. This type of system has been referred to as an advanced treatment unit or engineered pretreatment unit. These treatment processes may include effluent passing through a sand filter or an aerobic treatment unit. These systems aerate the effluent and change its composition through nitrification, leading to a decrease in NH_4^+ and increase in NO_3^-. While nitrification typically occurs in the STU for a conventional system, alternating aerobic and anaerobic conditions prior to discharge to the STU can be utilized to reduce N mass loading compared to STE alone. However, organic carbon must be consumed in the process, and it is unclear how this reduction in carbon may influence the treatment efficiency of the STU, or influence the transport of other constituents, such as microbial pollutants or

OWCs. Wastewater characteristics include the N concentration, pH (acidity) of the effluent, and carbon content. The pH of STE in the reviewed experiments was around pH 7, whereas the optimum pH for denitrification is 7.0-8.0 (Martin and Focht, 1977). An organic carbon source in the STE or STU is required for the process of denitrification.

2.3.2.6 Wastewater Source

The sources included in the literature review varied from single family residential homes to multi-family residential homes, seasonal campsites, and one cattle manure source. The source is an important parameter as the wastewater from domestic sources is expected to be less concentrated relative to office wastewater, because it includes gray water from laundry and other sources compared to the predominance of toilet use in an office.

2.3.2.7 Hydraulic Loading Rate (HLR)

HLR is directly associated with the wastewater applied to the STU, but it is also an indicator of contaminant mass flux. High flux of contaminants to the soil can reduce the efficiency of contaminant removal and increase the chance of the contaminants reaching groundwater. Other related parameters are the STU dimensions and geometry; the larger the STU infiltrative surface, the lower the expected contaminant load on the soil for a given total wastewater volume applied. Loading rate also can impact the saturation of the soil, which in turn influences transport and transformation of N (as described above) and other wastewater pollutants (discussed in subsequent sections).

The HLR can be expressed as volume of effluent applied to an area of soil per time, or as length units per time. HLRs in the reviewed sources were typically reported as cm/day, L/m^2/day, or gallon/foot^2/day. For consistency, the units chosen for this report are cm/day. When HLRs were reported as volume of effluent per day, the loading rates in cm per day were calculated by dividing the reported HLR by the STU dimensions. For example, a loading rate of 600 L/day equals 600,000 cm^3/day. If the infiltrative surface of the STU is 30 meters by 15 meters, than the STU infiltrative area equals 450 square meters, or 4.5 million square cm. The HLR in cm/day was then calculated as follows:

$$600{,}000\frac{cm^3}{day}/4{,}500{,}000cm^2 = 0.13\frac{cm}{day}.$$

HLRs were not always reported, especially for field experiments. For experimental studies, the researcher often controls the amount of STE applied to the soil each day, whereas HLRs under actual operating conditions may vary as a result of daily or weekly changes in water use. Most researchers conducting field experiments assumed that the amount of STE entering the soil was equal to the amount of water entering the source before the septic tank (see Table 2-2 for a list of effluent sources).

Missing HLR values were estimated based on a statistical analysis completed by McCray et al. (2005). In their work, the median value for domestic wastewater flow rate was found to be 227 liters per day per capita. This value was converted to cm/day, as explained above when the STU dimensions were known. Overall, HLRs were estimated for five experiments, and four other experiments were not matched with a HLR (one that used STE from a seasonal recreational facility, one that did not report the source of the STE, and two that did not specify STU dimensions).

2.3.3 Nitrogen Removal in Soil: Data Analysis and Discussion

The overall goal of this project is to develop simple tools that are useful for predicting N treatment with depth below the infiltrative surface. Statistically based empirical relationships are one type of tool that is being evaluated. In particular, it was hypothesized that literature data could be used to develop simple statistical relationships to predict N removal with depth based on other factors that are important for N transformation and that can be readily measured or estimated. This could enable a practicioner, for example, to estimate the expected treatment performance with depth by knowing only soil type, or HLR. To this end, a preliminary assessment of N removal in an STU under the influence of various factors was conducted. All 203 data points from the 84 experiments were plotted as C/C_o of total N vs. depth (Figure 2-4). As expected, there were more data points for shallow measurements than for deep measurements. With such sparse depth-dependent data, and because most studies reported a single N concentration value at one depth, it was difficult to obtain reliable conclusions about attenuation of N in soils with depth. Although the lowest ratios of C/C_o were seen at the greatest depths, no clear statistical relationship can be defined between removal and depth alone based on the reported literature values. Indeed depth and N removal were poorly correlated ($R^2 = -0.2$). This is not surprising given that many factors other than depth are known to be important factors in N removal (Section 2.3.2).

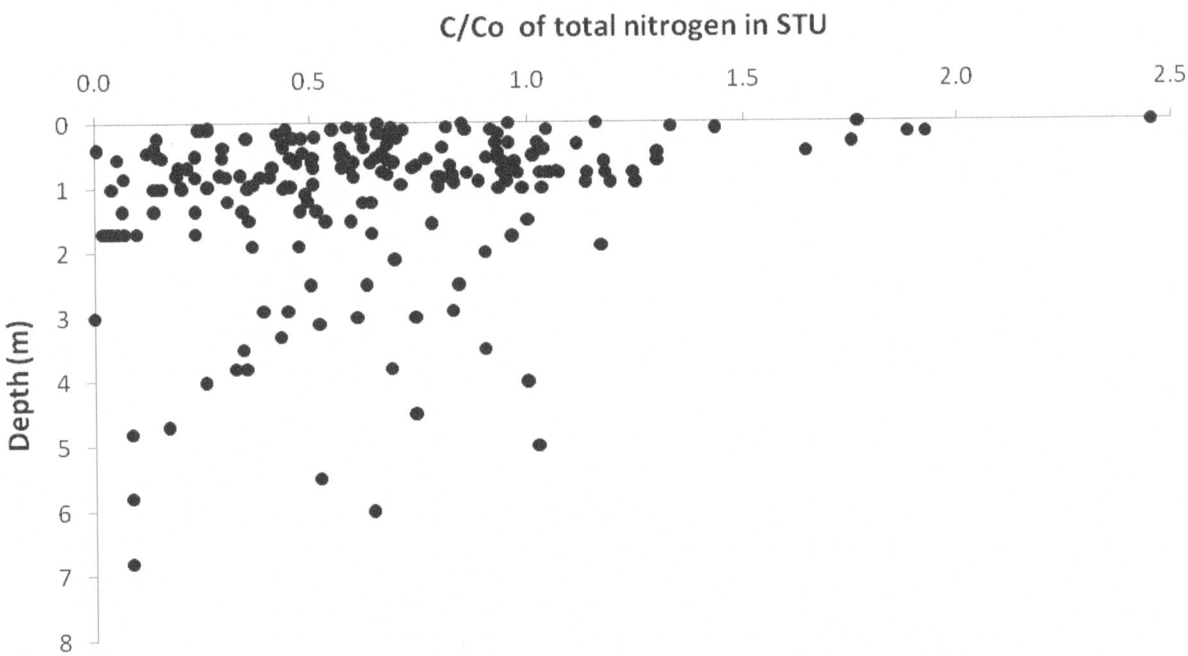

Figure 2-4. Summary of N Attenuation in STUs (all experiments shown).
C/C_o refers to the ratio of total N measured in the soil to the total N measured in the applied effluent.

Another important observation was that the N concentration in soil pore water could be higher than the concentration in the applied wastewater (see points with $C/C_o > 1$ in Figure 2-4). Because of the relatively large numbers of C/C_o values that are greater than one, it is not likely that this is due to analytical or reporting error. Possible explanations for this observation might be evapotranspiration (which does not affect the mass of total N, but increases the concentration), atmospheric deposition of N, seasonal changes in effluent concentrations, a buildup of sorbed NH_4^+ near the surface (and consequent increase in the concentration of dissolved ammonium in the soil solution), and alterations to the conditions for microorganisms in the soil.

Because all the data points in Figure 2-4 are from experiments that are quite different from each other, and because N transformations are theoretically known to be influenced by many different factors, further categorization of the data was attempted to examine potential relationships between N attenuation and depth for specific conditions, such as soil type and HLR. Figures 2-5 and 2-6 show the data in groups of different soil textural types. These graphs present examples for two contrasting situations: sandy soils (Figure 2-5) where many data points were available (n = 115), and organic soils (Figure 2-6) where only five data points were available. Such categorization is reasonable considering the different hydraulic, physical, chemical, and biological properties of different soils. The figures show that there is a clear enhancement in N removal with depth. However, no statistically meaningful predictive relationship between removal and depth exists for the different soil types (including soil types not shown here). Because soil texture is not represented numerically, initial correlations between N removal and soil type were estimated based on estimated values of K_{sat}, available water content, and the van Genuchten capillary parameters (van Genuchten, 1980). For each of these scenariors, no reliable relationship was observed ($R^2 < 0.2$).

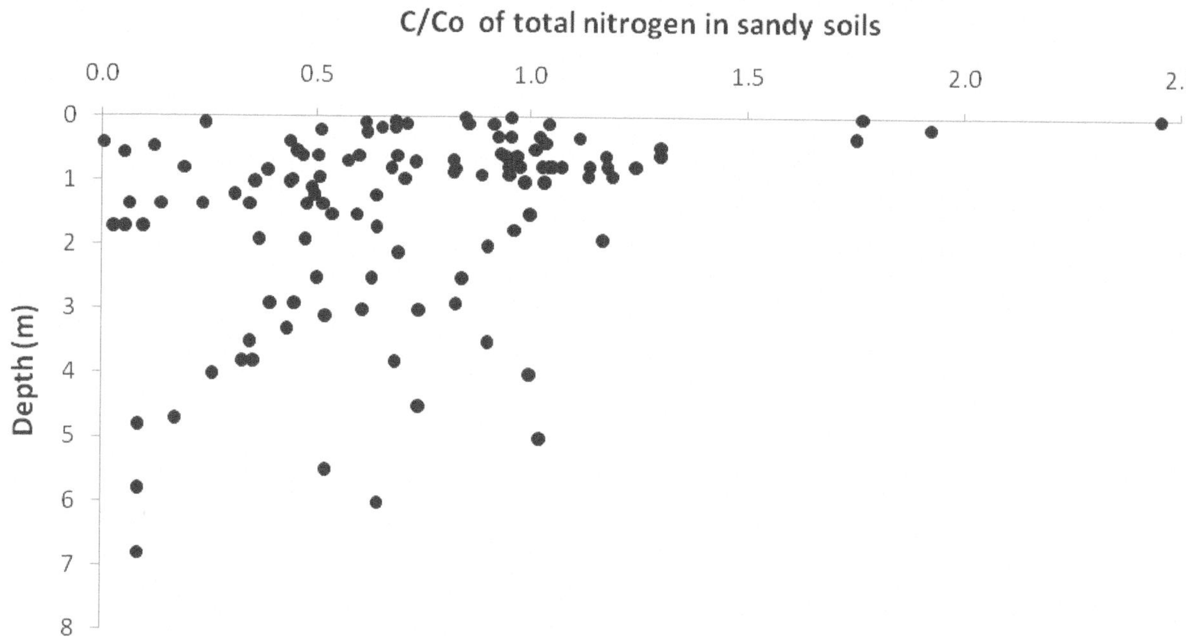

Figure 2-5. Attenuation of N in Sandy Soils.

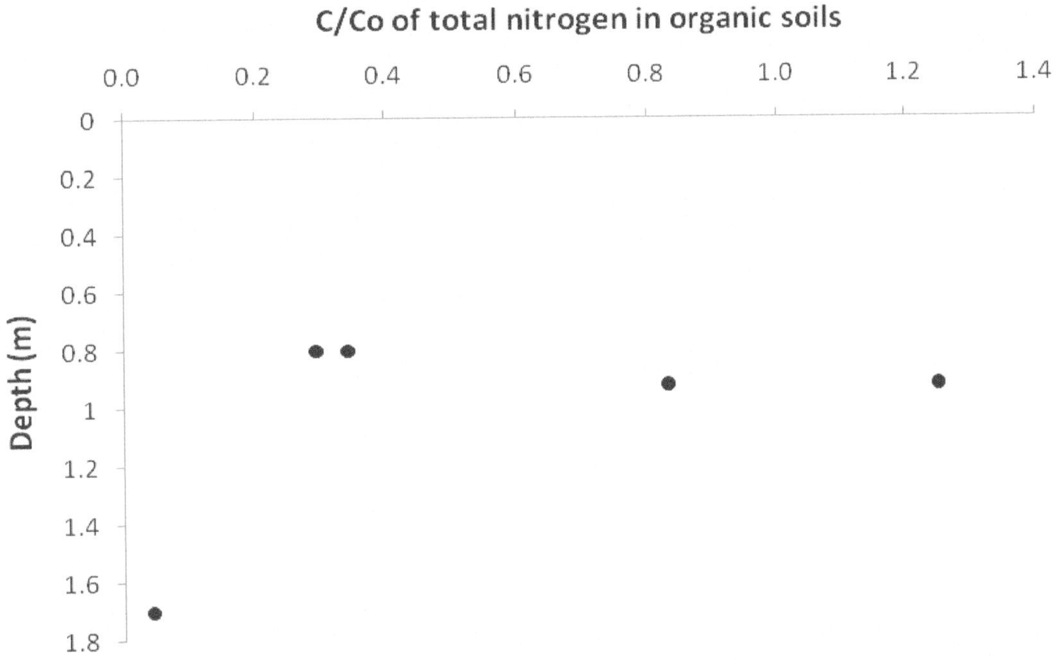

Figure 2-6. Attenuation of N in Organic Soils.

In addition to depth and soil texture type, it was expected that HLR could impact treatment (Section 2.3.2.7). However, HLR and N removal were again poorly correlated (R^2 = 0.40) when all data were utilized. Theoretically, depth, HLR, and soil type all are likely to influence treatment. Thus, the data were examined for empirical relationships that might predict N removal with depth on the basis of both soil type and HLR. C/C_o versus depth plots were chosen that separated soil type and incorporated HLR into the empirical equation (not shown on the plot). Based on this simple approach, no statistical relationship could be found for N attenuation in most of the soil types used. For clay loam, however, an empirical relationship was found between N attenuation, soil depth and HLR (Figures 2-7 and 2-8). Figure 2-7 shows the data (filled circles) along with the predictive empirical relationship (solid line). The empirical relationship is:

$$\frac{C}{C_o} = -0.09812\left(\frac{1}{Depth}\right)^{0.2088} + 0.58909\exp\left(-0.12918HLR^{-2.01423}\right) \qquad (2.3.3\text{-}1)$$

where C is the concentration of total N in soil pore water, C_o is the total N concentration in the STE, *Depth* is the depth below the effluent application in which C was measured, and *HLR* is the HLR of the effluent. The empirical relationship for clay loam soils indicates that HLR may be more important to the degree of nitrogen attenuation than depth. Specifically, the multiplication constant and the power of the depth variable in the equation have a low value, which means that for any given depth the C/Co result will not change much.

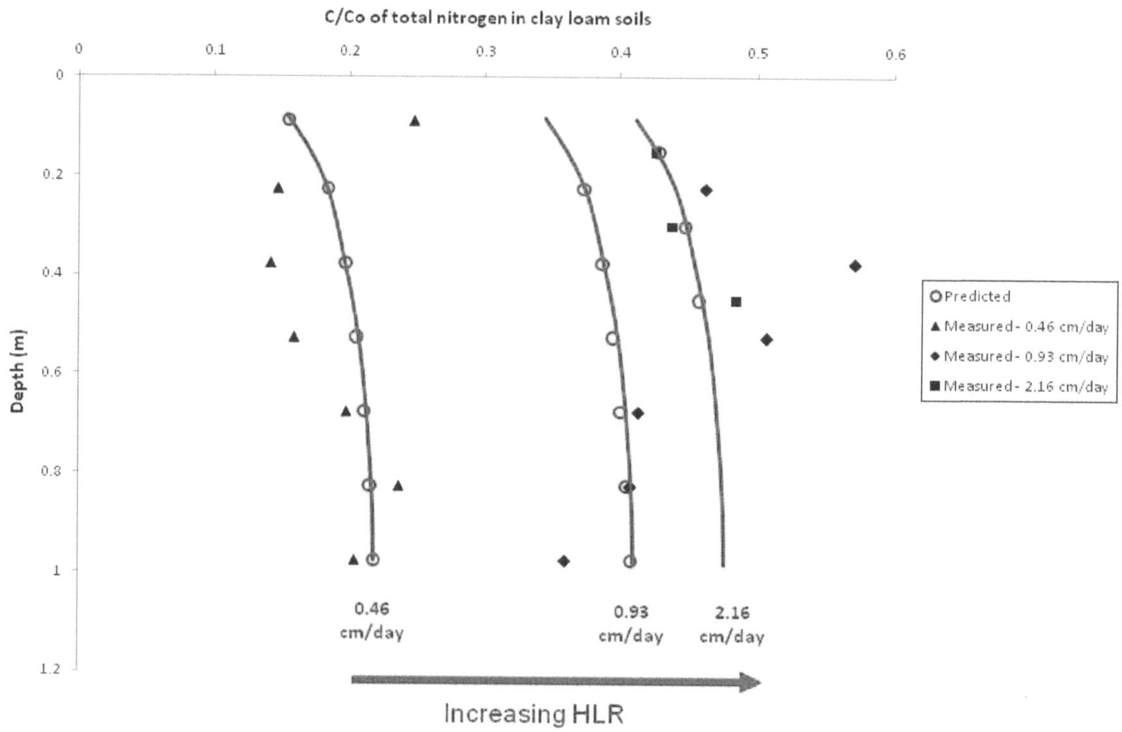

Figure 2-7. Attenuation of N in Clay Loam Soils with Depth.
C/C_o refers to the ratio of total N measured in the soil to the total N measured in the applied effluent.

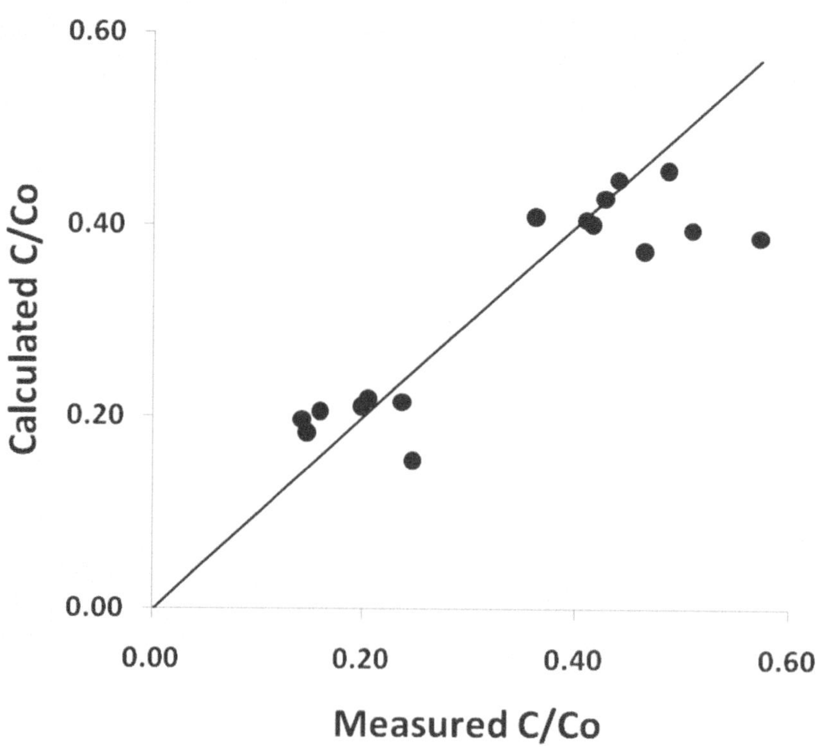

Figure 2-8. Comparison of Measured N Removal Values from the Literature and Predicted Values from Equation 2.3.3-1 in Clay Loam Soils.

Even though the data appear to be scattered, the predictive relationship that incorporates HLR fits the data quite well ($R^2 = 0.89$). The goodness of the prediction is more apparent in Figure 2-8, which shows the predicted (or calculated) values versus the data. The data for this clay loam texture relationship is based on three different independent experiments, with three different HLRs. While this relationship is promising, it demonstrates the complexity in the manner that different STU factors impact treatment. It also makes clear that simple binary relationships (e.g., correlation of C/C_o versus depth for various soil types) are not likely to produce statistically significant predictions, even when predictable relationships exist. Thus, multivariate statistical analyses are required to consider the effect that other factors (e.g., depth, soil texture, HLR, water content, and other soil conditions) and their interactions might have on N removal. Evaluating potential multivariate relationships requires a more rigorous and robust statistical approach than presented above.

ANOVA tests were conducted to determine if relationships between these selected factors could be identified. The ANOVA test attributes the variability of data to different factors. For instance, information derived from the ANOVA test can tell the relative degree that different factors affect the observation (if at all). In this case, the observation of N attenuation was evaluated for the factors of soil texture, soil depth and HLR. The results of the ANOVA general linear model test for soil depths of 30-60 cm ($n = 45$) attributed 66% of the variance in N attenuation to HLR, 20% to depth, 3% to soil textural class, and 11% to the variability within the data itself. This observation was similar to the empirical relationship for clay loam (eq. 2.3.3-1) and simple relationships where HLR correlated best to N removal compared to soil texture and depth. If this preliminary observation holds true after more rigorous data analysis, it would imply that wastewater application rate is a more important design factor than soil type for maximizing N treatment in an STU. However, we know from basic theory that soil type should play an important role in N treatment. Interestingly, ANOVA test results differed for data from laboratory studies versus data from field studies. For lab studies the soil type was the most important factor ($p < 0.001$), followed by depth ($p = 0.02$) followed by HLR ($p = 0.20$). For field studies the HLR was the most important factor ($p < 0.001$), followed by depth ($p < 0.001$) followed by HLR ($p = 0.08$). The lesser importance of soil parmeters in field studies is likely due to the great variability in soil properties, even within a particular soil type, at field sites. These results point to the the complexity in which different STU factors impact N treatment.

To enable tool development, other statistical analyses are also being investigated. Figure 2-9 demonstrates information that can be extracted from a technique called hierarchical cluster analysis (HCA) (Ward, 1963). In this case the Euclidean distance (similarity) between soil types or soil type clusters is described by:

$$d(i,k) = \sqrt{\sum_j (x_{ij} - x_{kj})^2} \qquad (2.3.3\text{-}2)$$

where *d(i,k)* is the distance between observations i and k. This type of approach enables the combination of soil types that have sparse data in a justifiable manner based on the HCA to ultimately develop more robust predictive statistical relationships. The grouping of soils into three or four categories in terms of HLR may be more practical given the highly imprecise hydraulic tests conducted in the field and the lack of scientific basis often employed during establishment of design HLRs used in regulatory codes (Lindbo et al., 2008; Siegrist, 2007). The HCA approach is useful to quantify and justify grouping categories.

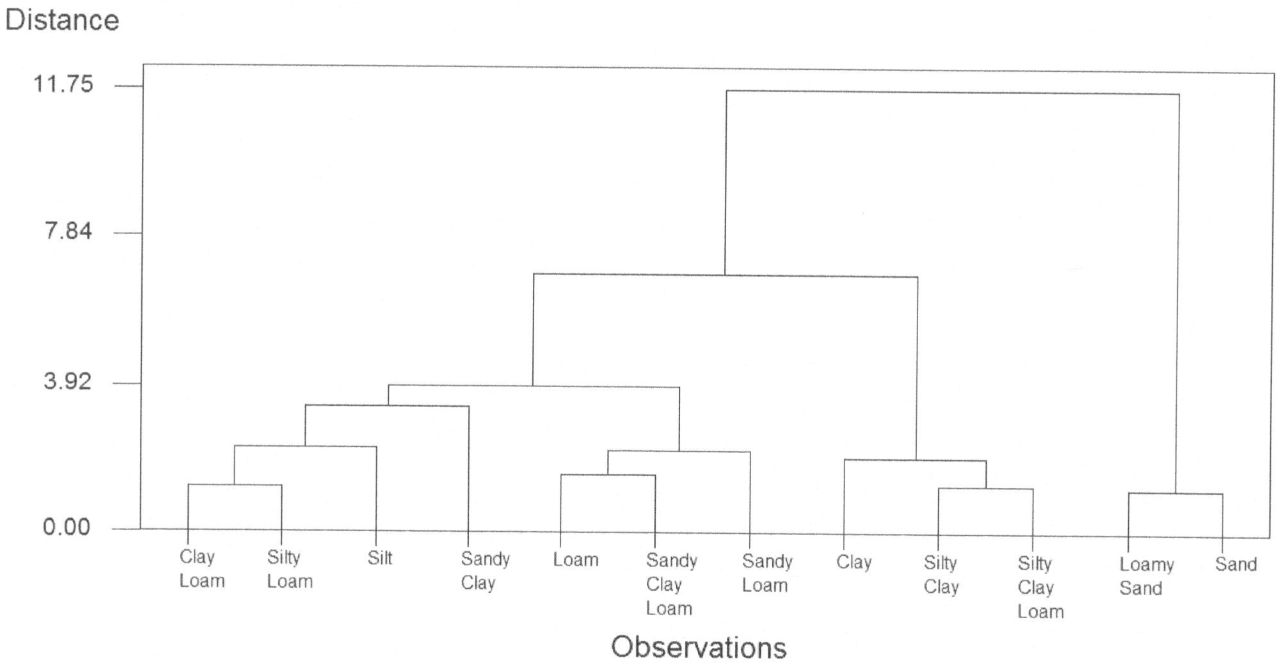

Figure 2-9. Cluster Test Results for Different Types of Soils based on Hydraulic Properties.
The Y axis represents the Euclidean distance (similarity) between soil types or soil type clusters (Equation 2.3.3-2).

Figure 2-9 suggests many potential groupings. For example, 4 groups could be formed as follows: 1) clay loam, silty loam, silt, and sandy loam; 2) loam, sandy clay loam, and sandy loam; 3) clay, silty clay, and silty clay loam; and 4) loamy sand and sand. Fewer groups can be identified by moving up on the HCA ladder illustrated in Figure 2-9. The HCA analysis with respect to N removal is being investigated in more detail.

As shown in Figure 2-3, and described earlier, N attention results were reported in the literature for many soil textures, but only limited data are available for most textures excluding sand and sandy loam. Yet it may be possible that different soils with different sand, silt and clay content do not vary much in their hydraulic or transport properties (e.g., saturated or unsaturated hydraulic conductivity, and soil suction versus water content relationships). Hydraulic properties are a function of soil type and theoretically known to be important with regard to treatment because they impact such factors as travel times, oxygen diffusion rates, and interphase partitioning. The soil texture triangle presented in Figure 2-2 defines soil by the sand, silt and clay content. While the change in any soil component is linear (from 0 to 100%), the associated change in soil hydraulic properties is not linear. For example, a slight increase in clay content can greatly reduce hydraulic conductivity. If certain soil types can be combined based on similar properties, then modified soil types that better incorporate common hydraulic properties can be created. For example, if it can be shown that silty clay and silty clay loam soils are similar with respect to hydraulic properties compared to other soil types, then development of a predictive relationship for N attenuation could be based on six experiments rather than three (Figure 2-3).

To determine whether two different soil textures could be assumed to share similar hydraulic properties, two computer software programs were used: The *Neural Network Prediction* feature in the HYDRUS-1D software (Šimůnek et al., 2005), which is coupled with the *Rosetta Dynamically Linked Library* (Schaap et al., 2001), and the Soil Water Characteristics software (Saxton and Rawls, 2004). The *Rosetta DLL* provides solutions to the van Genuchten equation (van Genuchten, 1980), which predicts unsaturated soil hydraulic conductivity and the soil moisture versus soil-suction relationships for different types of soils, either by soil type or sand/silt/clay ratios. The Saxton and Rawls (2004) method does not provide van Genuchten parameters, but predicts, for different soil types and sand/silt/clay ratios, several hydraulic properties, such as field capacity, wilting point, and hydraulic conductivity.

The parameters collected from the software were log-transformed and standardized (following methods of Güler et al., 2002), then Minitab software was used to conduct a HCA, with a Ward linkage method and a Euclidean distance measure (Ward, 1963). This method uses all available data to group different sub-categories into higher-level categories based on similarity of the chosen data, where the chosen data is the hydraulic parameters described above. The results of this analysis show that loamy sand, for example, is more similar to sand than to sandy loam based on these hydraulic parameters (Figure 2-9). Determinations like this could not be made based solely on the soil texture triangle. Because only four experiments were found which involve loamy sand, it is justifiable, based on this comparison, to combine them with the sandy soil experiments to achieve a higher number of experiments for the same category. In another approach, Radcliffe and West (2009) used the HYDRUS model to show that the 12 USDA soil textural classes could be grouped into four categories in terms of HLR.

Because important information related to the attenuation of N in soils is reported in the literature at various frequencies, additional statistical analyses will be conducted to develop tools that are easy to implement yet rely on rigorous experimental data and quantitative models rather than anecdotal observations. The statistical approaches described above and the importance of combining data sets, where justified, will enable further evaluation of incomplete data sets in support of tool-kit development.

2.3.4 Future Work

The factors that lead to denitrification, or removal of nitrates from the soil, are reasonably well understood. These factors include the presence of a carbon source, the ability of oxygen to diffuse to and from aerobic nitrification sites or anaerobic denitrification sites, wetness of the soil, suitable temperatures, and the transport velocity of nitrogen. However, most of these factors are not generally measured or reported in the literature (especially for field experiments). Fortunately, other parameters that are more easily measured may serve as surrogates for these important factors. Water content, for example, controls wetness and is directly (but nonlinearly) related to diffusion. Soil type in part determines porosity and tortuosity, which influences diffusion, and it is also related to organic matter content. The transport velocity is related to HLR, water content, and soil properties.

Some important findings have resulted from this preliminary analysis of literature data. First, from a practical perspective, HLR may be more important to N treatment within the first 30-60 cm than soil texture and soil depth. However, soil depth and soil texture remain important variables, although soil structure may prove to have a greater effect than soil texture. Univariate statistical relationships were not generally successful at yielding reliable empirical relationships to predict N removal versus depth for different soil types. An empirical bivariate relationship was developed for clay loam soils using both soil depth and HLR as the dependent variables. This result suggests that multivariate statistical relationships in general are useful and may be necessary to develop robust data based relationships. Additional efforts are being focused on these multivariate statistical relationships in the next phase of research. For scenarios where measured data are available, predictive statistical or empirical relationships are likely to be the most reliable predictive tools. However, given the variability and scarcity of data collected in field sites, it is unlikely that field data can be used to predict N attenuation for many relevant OWTS and STU operating conditions. Therefore, mathematical models are needed that incorporate relevant design variables and operating conditions. These models will be tested against actual data during the remainder of this project.

2.4 Phosphorus

Many studies describe phosphorus (P) transformations and sorption processes, mostly in regard to agricultural practices, yet only a few full-scale experiments have been conducted that relate to P in STUs. The primary concern with STU-related P is the transport from the STU to groundwater, and subsequently to surface waters. Special attention has been given to STUs that are in proximity to surface water environments (e.g., Zurawsky et al., 2004). A review of several studies found that the contribution of septic tanks to P loads in lakes can range from 4% to 55% of the total load (Lombardo, 2006).

2.4.1 Background

The motivation for P studies is the potential eutrophication (extensive algal growth) and disruption of the natural ecology in fresh water bodies caused by P release. Some nutrients, like N and P, occur at no more than micromolar levels and may be utilized almost to the point of exhaustion by the algae (Ryther and Dunstan, 1971). Field studies have shown that primary production in freshwater ecosystems is generally P-limited (Miettinen et al., 1997; Schindler, 1977), and P levels as low as 0.03 mg/L are associated with eutrophic lakes (Table 2-3) (Newton and Jarrell, 1999). An addition of P to fresh water may have two adverse effects: excessive algal growth, which leads to the deterioration of water quality and death of aquatic wildlife as a result of oxygen depletion; and a change in the nutrient balance in the water that favors growth of cyanobacteria (green-blue algae) (Schindler, 1977). When the N-to-P ratio (N:P) in a fresh water body becomes low (~5; Gerritse, 1993), due to an addition of P or depletion of N, algal competition for N is usually won by species that are capable of fixating N from the inert N_2 gas. Cyanobacteria have the N fixation capability, and also tend to release toxins to the water as secondary metabolism (metabolism by-product). Many cases have been reported of poisoning of vertebrate and invertebrate animals following ingestion of cyanobacterial bloom/scum material, and at least 60 different cyanobacterial toxins have been identified (Codd, 1995).

Table 2-3. Typical Total P Concentrations (mg/L) by Trophic State.
(Oligotrophic = lowest productivity, Hypertropic = highest productivity) Adapted from Newton and Jarrell, 1999.

	Trophic State			
Aquatic System	Oligotrophic	Mesotrophic	Eutrophic	Hypertrophic
Lake	<0.01	0.01 – 0.03	0.03 – 0.1	>0.1
River	---	<0.01	0.01 – 0.05	>0.05

Phosphorous is not usually considered a problem in drinking water. In Minnesota, for example, a minimum residual concentration of 1 mg/L of orthophosphate is maintained in water distribution systems to prevent corrosion and release of lead and copper from pipes (Rezania, 2004). On the other hand, Miettinen et al. (1997) sampled drinking water 10 km down the line from several waterworks plants in Finland, and observed microbial growth with an addition of as little as 0.01 mg/L phosphate (PO_4^{3-})-P to the samples (the background levels of the drinking water were <0.002 mg/L PO_4^{3-}-P). Excess intake of polyphosphates inhibits the adsorption of calcium at the intestine (Sekiguchi et al., 2000). However, these compounds are not a concern in STUs because they degrade to phosphates.

As it became clear that P release to the environment had adverse ecological effects, several States began to ban or partially ban the use of P in detergents. This trend started in the 1970's and continued to 1994, when the manufacture of phosphate detergent for household laundry was ended voluntarily by the industry (Litke, 1999). It should be noted that detergent bans and guidelines typically apply to phosphate for household laundry detergents only and do not apply to dishwashing detergents or commercial cleaning products (Litke, 1999). The percentage of U.S. households with dishwashers increased by 9.4% from 1992 to 2002, and the percentage of U.S. households with clothes washers increased by 5% at that same time (U.S. Census Bureau, 2005). The percentage of households with clothes washers was much higher than the households with dishwashers (80% compared to 58.1% in 1998), but considering the rise in dishwasher ownership, and the continual use of polyphosphate-containing dishwashing detergents, dishwashers have become an increasingly important source for P in wastewater.

Wastewater treatment plants are required to release their effluents with a typical P concentration upper limit of 0.5-1.5 mg/L, and the U.S. EPA has a recommended limit of no more than 0.1 mg/L of total-P in rivers (Litke, 1999). However, there is currently no regulation regarding allowed P concentration in STE or groundwater, or a set goal for P removal in STU. Instead, total maximum daily loads (TMDLs) have been implemented for many surface water bodies. A TMDL is a calculation of the maximum amount of pollutant that a water body can receive and still meet water quality standards, and an allocation of that amount to the pollutant's source (Lombardo, 2006). For example, a TMDL developed for Lake Carlton, FL, requires 41% reduction in non-point P sources, of which septic systems accounted for 14% (Lombardo, 2006, after Florida Department of Environmental Protection, 2003).

2.4.2 Phosphorus Chemistry

Phosphorus is commonly found in the following forms:

- Orthophosphate is an inorganic series, which includes H_3PO_4, $H_2PO_4^-$, HPO_4^{2-}, and PO_4^{3-} species. The transition between these species is pH dependent, with $H_2PO_4^-$ and HPO_4^{2-} the dominant orthophosphate species in neutral-pH waters.
- Organic P includes P incorporated with organic compounds, such as sugars, phospholipids, and nucleotides. Adenosine-triphosphate, or ATP, is a phosphorous-bearing organic molecule that serves as an energy storage unit in living cells. Some synthetic organic compounds, including some insecticides, contain reduced forms of P.
- Phosphorus minerals result from reactions of phosphate with different cations (mostly iron, aluminum, calcium, and lead) and lead to the formation of minerals such as hydroxyapatite, fluorapatite, strengite, and vivianite. The source of these minerals is usually magmatic rocks, yet they can be found in sedimentary environments where there is enough P and suitable cations (e.g., marine environments with high productivity, coastal environments enriched with bird droppings [guano]). Hydroxyapatite is also the major constituent of tooth enamel, which is the hard outer part of the tooth. Minerals that contain calcium and P (calcium phosphates) are commonly used as fertilizers.
- Condensed phosphates are derived mostly from detergents and cleansers, for water "softening" (removing of divalent cations such as calcium and magnesium to prevent precipitation of carbonates). They include polyphosphate forms such as $P_2O_7^{4-}$ and $P_3O_{10}^{5-}$,

and degrade, or hydrolyze, slowly to orthophosphate (see above). Polyphosphates are an additive in some public water supplies as a means of controlling corrosion (Rezania, 2004). They are also used widely as food additives in cheese, fish paste products, and ham and sausage to prevent discoloration and to stabilize vitamin C (Sekiguchi et al., 2000).

- Elemental P is present in several small-scale applications such as fireworks, flares, napalm and safety matches.

Unlike N, which usually undergoes an irreversible process of mineralization-nitrification-denitrification in the STU, P participates in processes which can be reversible. These processes include adsorption and precipitation. Adsorption is the binding to soil particle surface, as phosphate is "attached" to positively charged particles such as aluminum, iron and manganese oxides and hydroxides, and some clay particles, and is limited by the number of available sorption sites. Precipitation is the formation of a solid mineral in solution from a chemical reaction, and theoretically is not limited as long as sufficient aqueous concentrations of mineral components are available, and as long as there is space for the mineral to "grow". This process is often considered to be irreversible. However, the reversibility depends on several factors, such as pH, redox (reduction-oxidation) conditions and the solubility product of P-bearing minerals (Robertson et al., 1998). The solubility product of a specific mineral is a measure of the concentration of the mineral components (ions) in solution that is required to precipitate the mineral. The more ions available, the greater the chances are for precipitation and for removal of P from soil pore water. The complication in predicting P behavior in STU stems from the different environments that promote different processes. Adsorption and precipitation processes are described in more detail below.

2.4.2.1 Adsorption

As previously mentioned, the major forms of orthophosphate in wastewater are anions ($H_2PO_4^-$ and HPO_4^{2-}) which will be attracted to positively-charged mineral surfaces. Quartz sand, for instance, is a poor phosphate adsorbent because quartz is negatively charged at pH values of approximately 3 and higher (Sposito, 1989). Minerals that are positively charged at near-neutral pH typical in STUs include aluminum oxides and hydroxides (corundum, gibbsite, boehmite, allophone), iron oxides and hydroxides (goethite, ferrydrite, hematite), manganese oxides (pyrolusite), and variable-charge clay minerals. These minerals or oxides are positively charged because of the substitution of H^+ protons in the surface functional groups. In acidic solutions, where the amount of H^+ protons increases, the affinity of phosphates to adsorb to mineral surfaces increases.

Sorption and precipitation are often related. Phosphate can adsorb to about 5% of pure calcite (calcium carbonate) surfaces, despite the mineral's negative charge under slightly-basic to basic pH conditions. While only a small portion of the mineral's surface is available, these adsorption sites act as nuclei for the precipitation of calcium phosphate minerals, mostly at high P concentrations (Borrero et al., 1988). At low P concentrations, iron is still a more efficient adsorbent than calcium carbonate, whereas fine-grained calcium carbonate adsorbs P slightly more than clay minerals such as illite and smectite, due to a higher specific surface area (Borrero et al., 1998).

Organic matter causes contradicting effects on phosphate sorption. Organic acids can compete with phosphates for sites on mineral and soil surfaces, thereby decreasing phosphate adsorption. Certain organic acids can entrap reactive aluminum and iron in stable organic complexes called chelates, thus making them unavailable for reaction with P. On the other hand, phosphates can be immobilized by organic carbon, as organic-bound complexes, and thus removed from solution. Overall, organic soils are not efficient in removing phosphates (Brady and Weil, 2002).

Allophane is an aluminum silicate clay mineral with an extremely high P adsorption, due to its large surface area. It is typically found in soils associated with volcanic ash (Brady and Weil, 2002).

Brady and Weil (2002) summarize different soil components' capacity of removing phosphorous in the following order (from highest to lowest):

- Amorphous Al, Fe and Mn oxides; allophone;
- Crystalline Al, Fe and Mn oxides;
- Carbonate minerals;
- 1:1 clays (e.g., kaolinite); and
- 2:1 clays (e.g., montmorilonite).

2.4.2.2 Precipitation

In calcareous soils and marine coastal environments, the most common P mineral precipitate is apatite ([$Ca_{10}(PO_4)_6X_2$], where X is an anion, mainly hydroxyl [OH^-] and fluoride [F^-]). Typical calcium sources for the precipitation reactions are: shell fragments in coastal marine environments (Whelan, 1988); in semi-aridic and aridic terrains a calcareous horizon is typically developed in the soil (Sposito, 1989); calcium carbonate can precipitate as a normal weathering product of calcium-bearing primary silicates, such as pyroxene, amphiboles and feldspars (Sposito, 1989); and calcium can also be found in soils developed from carbonate rocks (Borrero et al., 1988). Acidic pH encourages dissolution of calcium-carbonate and release of calcium ions into solution, which should theoretically contribute to the formation of calcium phosphate minerals. However, hydroxyapatite [$Ca_{10}(PO_4)_6(OH)_2$] is more stable under basic conditions (Maurer et al., 1999).

Phosphate tends to bind to calcium cations in basic soils with high Ca/P ratio, but to iron, aluminum, magnesium and manganese cations in acidic soils. In non-calcareous terrain, the development of acidic conditions cause gibbsite [$Al(OH)_3$] dissolution and, subsequently, variscite [$AlPO_4 \cdot 2H_2O$] precipitation (Zurawski et al., 2004). Calcium phosphate minerals such as hydroxyapatite [$Ca_{10}(PO_4)_6(OH)_2$] and fluorapatite [$Ca_{10}(PO_4)_6F_2$] have lower solubility products (i.e., they are less likely to dissolve and release phosphates back to the solution) than iron phosphate minerals such as strengite [$FePO_4 \cdot 2H_2O$] and vivianite [$Fe_3(PO_4)_2 \cdot 8H_2O$]. The formation of struvite [$MgNH_4PO_4$], for example, in wastewater is improbable due to its large solubility product and the fact that its equilibrium is rapidly achieved (Maurer et al., 1999). A solution can be supersaturated with ions that are expected to form minerals, due to a slow rate of mineral formation (Robertson et al., 1998).

It is believed that precipitation of phosphate minerals such as hydroxyapatite (HAP) is a two-step process. First, a surface complex is formed, which is not a pure compound under "dirty waste-water conditions" but an amorphous mixed crystal with a higher solubility product than HAP (Maurer et al., 1999). This surface complex has a relatively large solubility product. Second, over time the surface complex exothermically crystallizes into the less soluble HAP. There is a link between aerobic/anaerobic environments and the stability of phosphate minerals. In anaerobic (reducing) environments, the dominant iron form is the soluble Fe^{2+} (ferrous ion), whereas in oxidized environments the dominant form is the insoluble Fe^{3+} (ferric ion). Vivianite, which is composed of ferrous ions and phosphates, is a major P sink under reducing environments, with a lower solubility product under basic conditions (Robertson et al., 1998). In oxidizing environments the dominant iron phosphates are variscite or strengite, which have lower solubility products under acidic conditions (Brady and Weil, 2002). Under field conditions, precipitation of phosphate minerals should be considered a kinetic process.

2.4.2.3 Materials that Enhance Phosphorus Removal

Several studies suggested practices for removal of P from wastewater in STUs. These studies included sorption of phosphates or encouragement of phosphate precipitation.

Bellier et al. (2006) conducted batch tests, laboratory lysimeter experiments and field-scale experiments which examined P attenuation from wastewater in columns filled with apatite-containing materials. The presence of the crystal acts like a catalyst which lowers the activation energy barrier between the adsorbed calcium phosphate and the nucleation/crystallization of additional apatite. In their study, Bellier et al. tested the P retention capacity of apatite minerals from sedimentary and magmatic sources. They found that adsorption occurred almost instantaneously and was rapidly followed by precipitation. Higher retention of P was achieved in columns filled with sedimentary apatites. Because adsorption is a surface phenomenon, it is considered that the high densities associated with igneous materials were not favorable for igneous apatites to act as adsorbents. The high affinity found for sedimentary apatite could be explained by the lower density compared to the igneous apatites. In lysimeter columns, the best retention results were achieved for a 50/50 mixture of limestone and sedimentary apatite. The apatite encouraged nucleation, and the limestone dissolution provided free calcium ions to react with phosphates (it is possible to create pure HAP crystals with Ca:P ratios of 1.3:1 to 2:1; Maurer et al., 1999). In this test, there was 100% P retention during the first 15 days, and approximately 60% retention after 39 days. In the field experiment, there was about 60% decrease in total-P between wastewater (7.6 mg/L) and column effluent (1.9 mg/L) concentrations.

Brandes et al. (1974) examined P removal in filter beds applied with STE, during time frames of several weeks to about a year. They used 10 filter beds with different materials and grain size distributions. The least amount of total-P removal, 1% and 4%, was observed in columns with coarse sand, D_{10} = 1 mm and D_{10} = 2.5 mm, respectively (D_{10} is the particle diameter when the percentage less than of the particle-size distribution curve is 10%, by weight). The best total-P removal, 88%, was observed in a column with 15 in. depth of sand (D_{10} = 0.24 mm), overlaying a 15 in. mixture of 10% red clay and 90% sand. High removal rates (73-74%) were observed also for columns with a mixture of sand and limestone, and sand with equal amount of clay. 76% removal was observed at a column with 22 in. depth of sand (D_{10} = 0.24 mm) overlaying a 8 in. mixture of 4% red clay and 96% sand.

Some experiments reported in the literature used relatively low-cost materials for removal of phosphates:

- Flyash is a by-product of coal-burning power plants. Vinyard and Bates (1979) used a high-calcium fly ash for phosphate removal from secondary-quality wastewater. In several experiments, they mixed various amounts of flyash with 1 liter of wastewater. They reported 98% P removal within 5 minutes, from an initial P concentration of 5.6 mg/L, with a flyash concentration of 25 g/L. If one assumes a domestic wastewater volume of 520 liters per day

(Tucholke, unpublished data), with a 5.6 mg/L concentration of P, then the amount of flyash required for similar P removal would be approximately 4.7 tons per year.

- Tamarind nut shell activated carbon (TNSAC) is a by-product from the processing of the tamarind seed nut and used as an additive for cattle feed. Bhargava and Sheldarkar (1993) reported 95% removal of phosphates by un-rinsed TNSAC, with an initial phosphate concentration of 55 mg/L. The TNSAC reached phosphate saturation after about 1 hour, and the TNSAC adsorption capacity was 8.7 mg PO_4^{3-}/g TNSAC. Natural soils that remove more than 0.35 mg P/g soil (1.07 mg PO_4^{3-}/g soil) from solution are generally considered to be high P-fixing soils (Brady and Weil, 2002).
- Slag is a by-product in iron ore with caustic lime refining process. Hisashi et al., (1986) found that slag has the capability of removing about 9 mg PO_4^{3-} per g slag (initial phosphate concentration: 400 mg/L), with optimum adsorption at pH 8. They attributed the high sorption capacity of the slag to its high porosity and to its constituents: iron oxides, aluminum oxides and hydroxides, calcium carbonates and hydroxides, and magnesium hydroxides.

2.4.3 Phosphorus Fate and Transport in Soil Treatment Units

Common phosphorus forms in domestic wastewater are orthophosphate, polyphosphate and organic P. Between 20 and 30% of total P in raw wastewater is separated out in the form of sludge in the septic tank (Lombardo, 2006, after Wood, 1993). However, current monitoring of raw wastewater and STE from 17 single-family residences indicates little attenuation of P in the septic tank (personal communication with Maria Tucholke, CSM, unpublished data from WERF 04-DEC-1, Lowe, et al.). Some of the P probably precipitates in the septic tank with iron to form the mineral vivianite (Robertson, 1998).

Rothe (2006) found that the median and average total P concentrations in a single-source-domestic STE are 10 mg/L and 12.2 mg/L, respectively. His study included 46 reported values, with a concentration range of 3-39.5 mg/L. For multi-source domestic STE, the median and average values were 6.9 and 7.03 mg/L, respectively, based on 6 reported values. However, Rothe indicated that many of his sources dated from before the phosphate laundry detergent ban. McCray et al. (2005) report a median phosphate-P concentration of 9 mg/L in STE, based on 35 data sources, with a 1st and 3rd quartile values of 6 and 13 mg/L, respectively. Crites and Tchobanoglous (1998) report a typical total P concentration of 16 mg/L in domestic STE. Of that concentration, 10 mg/L is inorganic P, and 6 mg/L is organic P.

2.4.3.1 Phosphorus Attenuation in Different Soil Treatment Unit Conditions

The importance of a biological film (clogging zone) at the infiltrative surface in trenches has been noted in the past (Siegrist, 1987; Magdoff et al., 1974). This wastewater-induced bio-film contributes to the STU in the following ways: it promotes a spatially uniform application of STE throughout the STU by causing ponding in the trenches; it helps maintain unsaturated conditions in the soil underneath the trench; it increases the soil biogeochemical activity; and it can enhance sorption, biotransformation and die-off/inactivation processes (Siegrist, 1987). Magdoff et al. (1974) applied wastewater to test columns, and observed a decrease in P concentrations once ponding conditions started to occur in the columns. The P concentrations in their sandy loam soil under bio-film conditions were stabilized at 2-6 mg/L, compared to 11-14 mg/L in columns with no bio-film. The total P removal in the bio-film columns was 91%.

Ver Hey and Woessner (1987) conducted a field STU experiment in two single-family homes in Minnesota, with six-year-old septic systems. They found that no biomat formed at the infiltrative surface, which led to high percolation rates in the loamy soil. They also found, based on several spatial sampling points, that most of the effluent was discharged at the STU near the inlet, which meant that only a small portion of the STU was actually utilized for treatment. The amount of effluent released from the septic tanks did not exceed regulations (200 gallons per acre per day, compared to the 600 gallons per acre per day regulation), but the poor distribution over the STU suggested a much higher actual load to the effective STU. The orthophosphate measurements showed virtually no treatment in the STU, with reduction of orthophosphate only in the groundwater, probably due to dilution.

Cogger et al. (1988) support the idea that higher loading rates lead to increased leaching of phosphates to groundwater. In their experiment, they sampled groundwater 0.5 m down-gradient from STU trenches with different loading rates (1, 4 and 16 cm/day) in Virginia. They found that higher loading rates of STE resulted in a higher concentration of soluble P in groundwater. They attributed this relation to a higher soil pore-water flow velocity. As the flow velocity increases, there is less time for equilibration to occur between soil particles and P. There is also less time for complete adsorption/precipitation to occur. As a result, less P is retained at higher velocities. Their findings seem to support findings of other researchers (de Camargo et al., 1979; Lance and Gerba, 1977), that there is a difference between fast and slow interactions of P in the soil. The fast interactions (kinetic sorption or precipitation) are independent of flow velocity which initially is the most important retention mechanism. The slower velocity-dependent reactions (equilibrium precipitation or sorption) are more effective removal processes for slow soil-water velocities. More observations from Cogger et al. (1988) support the role of flow velocity in P leaching. In their study, some of the filtration beds were flooded with groundwater during winter, yet there was no flush of P into the groundwater. In contrast, groundwater down-gradient from the non-flooded filtration beds showed increased P levels during heavy rains. This suggests that desorption (the opposite of adsorption) is greatly enhanced by flow velocity changes, such as during heavy fall and winter rains. The levels of P in groundwater declined as the soil became drier.

Zurawsky et al. (2004) studied mature STUs (23 and 44 years old) in calcareous and slightly calcareous sandy soils in Canada that have operated with no decline in redox conditions since their installation. When P is retained in association with Fe(III) solids, a situation that appears common below septic system tile beds, vulnerability to P remobilization appears to exist should redox conditions change (Zurawsky et al., 2004). Systems that experience an increase in STE loading may have an the increase of the degree of water saturation in the STU, which, in addition to increasing dissolved OM and NH_4^+ loading, increases the likelihood that reducing conditions will develop. Results of their column tests, which included soil samples from the STUs, suggest that reducing conditions could lead to remobilization of previously immobilized P. An example of systems that are vulnerable to such change in loading rates is lakeshore cottages designed for seasonal use that are converted to permanent dwellings. It should be noted, that despite system stability in both of the sites in the study, distinct P-rich plums exist in groundwater underneath the STUs, with P concentrations of 1.3 and 4.9 mg/L (sampled in proximal plume cores; Robertson et al., 1998).

Doyle et al. (2005) measured STE components that can precipitate and form phosphate minerals. They found, based on STE constituents and saturation indexes of different minerals, that precipitation of hydroxyapatite is very likely immediately after STE is release to the soil. With further infiltration of wastewater into the soil, the soil pore water becomes less saturated with phosphates and calcium, and adsorption becomes a more important attenuation mechanism of P until the wastewater reaches the water table. Doyle (2006) also found that P attenuation in STUs cannot be explained by equilibrium processes alone, based on comparisons of batch tests with field scale data. This suggests that kinetic precipitation, and possibly kinetic desorption, occur in STUs – phenomena that are much harder to model than equilibrium sorption and precipitation.

2.4.3.2 Phosphorus Attenuation with Depth

Only 11 sources describing 30 experiments were found in the literature which report P concentrations with depth in STUs. The distinction of one experiment from another is by a change of conditions, such as loading rates, different location (for field experiments), and soil types. Unfortunately, 27 of the 30 experiments reported only a single value of P concentration with depth. In some of the articles the P concentration was reported as total P, in some it was reported as PO_4^{3-}-P concentration, and in some it was mentioned as PO_4^{3-}. A summary of the experiments' results, as C/C_o (ratio of concentration in soil pore water and concentration in the STE) versus depth, are presented in Figure 2-10.

Figure 2-10a is consistent with the findings of Zurawski et al. (2004), who found that the first ~0.5 m below the infiltrative surface, which they called the "High-P zone", were enriched with retained P compounds. In their study, concentrations of P deeper than 0.5 m were quite low (but still higher than background levels). P concentrations were determined based on the amount of P and Fe compounds that leached from samples, taken from different depths in the STU, when redox conditions changed. Figure 2-10a suggests that, based on several experiments, total P concentration in soil pore water decreases drastically about 1 m below the infiltrative surface. However, it is not known if this trend is meaningful or is simply due to the sparse data points below 1 m.

Figure 2-10b, shows a similar concept but for phosphate-reported concentrations. In this graph, a greater scatter in the reported values at depths below 1 m is observed. Again, it is difficult to be conclusive about P attenuation based on this plot because most of the data points below 1.5 m are taken from only two experiments reported in the same reference. One of these experiments measured STE discharge from a soak-well rather than a conventional STU. Under soak-well conditions we can assume a continuous flow, which would result in increased P levels in the soil, as was observed in the experiment.

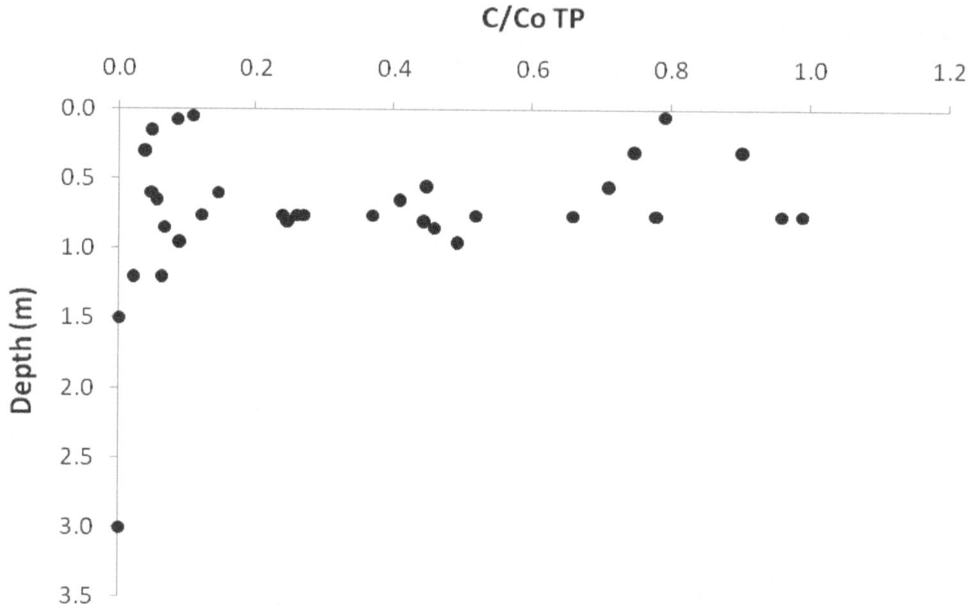

a)

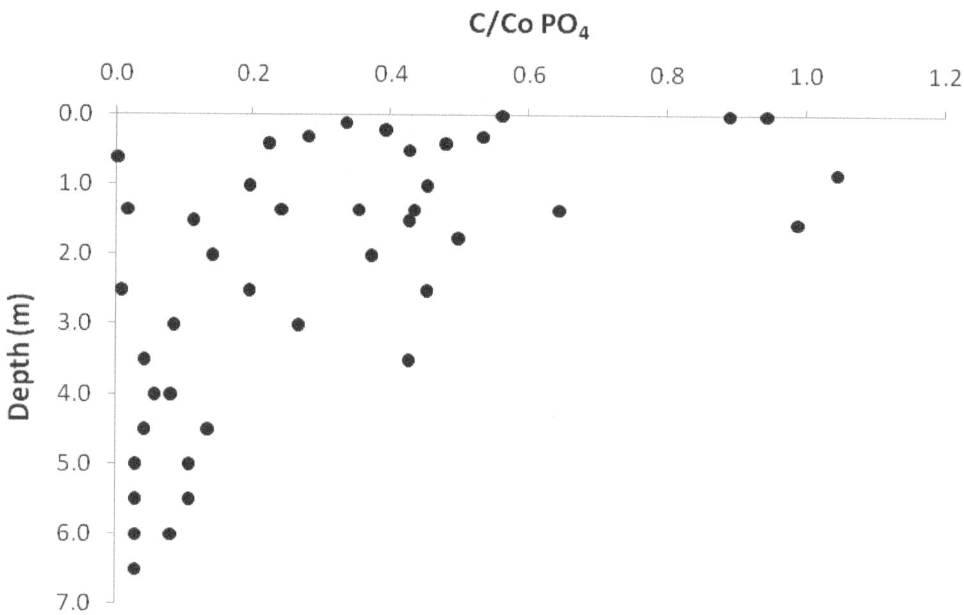

b)

Figure 2-10. Attenuation of (a) Total P with Depth in STU Pore Water, and (b) Attenuation of PO_4^{3-} with Depth in STU Pore Water.

It is important to remember that the plots demonstrate P attenuation in STU pore water (i.e., soluble P). The amount of all P forms in the STU (soluble, mineral and organic) does not necessarily decrease with time. In fact, crystallization of P minerals indicates removal of soluble P from soil pore water and minimization of the risk of P leaching into groundwater, as long as conditions do not change drastically in the STU. Unlike inorganic N compounds, which can be theoretically removed from the soil by denitrification, P that is not leached into groundwater accumulates in the STU and is effectively removed from the infiltrating wastewater.

2.4.4 Conclusions

STE-derived P is involved in several chemical processes in the STU, including degradation of polyphosphates to orthophosphates, adsorption and precipitation. Adsorption of orthophosphates is usually considered a two-phase reaction (with an instantaneous/equilibrium phase and a kinetic phase) in the STU, and depends on available sorption sites and soil constituents, with iron and aluminum oxides and hydroxides having the highest capability to adsorb. Sorption is also pH and redox-state dependent. Precipitation is considered to occur in several steps, with different kinetic rates (e.g., Doyle, 2006; Enfield et al., 1981), and also is dependent on pH, soil constituents, aqueous concentrations of P and other ions, saturation of the soil, solubility constants, and flow rate of wastewater in the STU.

Most of the P-soil interaction data found in the literature appear to be in the form of batch tests. These are meant to determine primarily equilibrium adsorption capabilities of the soil. However, these experiments do not investigate kinetic behavior, nor provide long-term projections for precipitation of phosphate compounds, which is a much slower process than adsorption. A few studies explored long-running septic systems with regard to P behavior in STUs (e.g., Robertson et al., 1998; Zuwarsky et al., 2004). These show that with time, P plumes can develop in groundwater under STUs, even if the systems appear to be working properly. High HLRs or uneven distribution of STE in the STU contribute to P leaching from STU, as well as lack of a clogging zone across the infiltrative surface of the STU trenches.

However, not enough information exists in the literature to determine the P retention in the STU based on soil type, mostly because P attenuation processes are not fully understood and hard to model. Without a regulatory limit on the amount of P allowed in groundwater, it is challenging to accurately predict P removal in an STU, and may require multivariate statistical analysis on the data collected in this review.

2.5 Microorganisms

2.5.1 Background

A wide range of human-pathogenic microorganisms, which include enteric viruses, enteropathogenic bacteria, and protozoa are found in STE (Table 2-4). In soil-based OWTS, the STU is expected to remove and/or inactivate these biological contaminants, preventing their transport to surface or groundwater. Removal and inactivation of pathogens depends on system design parameters and soil properties and the dynamic interaction between these.

Among the most important OWTS design parameters are the distance to groundwater or surface water, HLRs, and/or dosing schedules. Most onsite wastewater codes require a minimum separation distance of 45 cm between the infiltration zone of the STU and the seasonal high water table or saturated zone, irrespective of soil characteristics (U.S. EPA, 2002a). However, the effectiveness of these separation distances can be significantly influenced by the loading rate and dosing pattern, as well as by soil properties, and are not necessarily designed for optimal removal of all types of pathogens. Soil properties considered to be important in pathogen removal include texture, structure, pH, temperature, and the presence of a biomat and its associated biotic community (Canter and Knox, 1985; Gerba and Goyal, 1985; U.S. EPA, 2002a). The biomat is a layer of organic and inorganic debris that includes bacteria and their excreted extracellular polymers (Ronner and Wong, 1994). This layer accumulates at the soil/STE interface and is approximately 2.5 cm thick (U.S. EPA, 2002a). As the biomat develops, it forms a clogging layer in the soil, which slows infiltration of STE, allowing for unsaturated conditions to be maintained below (Van Cuyk and Siegrist, 2001). These properties interact with STE, controlling infiltration, particle retention, oxygen availability and degree of soil saturation, all of which are considered important in pathogen removal.

Table 2-4. Concentration of Selected Pathogens Found in STE and their Infectious Dose.

Pathogen	# in STE (per 100 ml)	Infectious dose (# of organisms)
Enteric viruses	$0\text{-}10^{7}$[a]	1[d]
Enteropathogenic *E. coli*	$10^{5}\text{-}10^{8}$[a]	10 for O157:H7[b] $10^{6}\text{-}10^{7}$ for other species
Clostridium perfringens	$10^{3}\text{-}10^{5}$[c]	10^{8}[b]
Pseudomonas aeruginosa	$10^{3}\text{-}10^{4}$[c]	
Salmonellae	$0\text{-}10^{7}$[a]	15-20[b]
Shigellae	$0\text{-}10^{7}$[a]	10[b]
Vibrio cholerae	$0\text{-}10^{7}$[a]	10^{6}[b]
Giardia lamblia	$10^{3}\text{-}10^{4}$	1[b]
Cryptosporidium parvum	$10^{1}\text{-}10^{3}$	1[b]
Entameoba histolytica	$0\text{-}10^{5}$[a]	1[b]

[a] (Feachem et al., 1980)
[b] (US Food and Drug Administration, 1992)
[c] (Crites and Tchobanoglous, 1998) number is reported for STE and raw wastewater
[d] (Bitton, 1980)

This chapter summarizes the literature on removal and inactivation of three groups of pathogenic microorganisms found in STE (viruses, bacteria, and protozoa) and evaluates the mechanisms of removal . Those properties of soil-based treatment systems that most influence microorganism removal are also identified. The discussion emphasizes factors that can be manipulated through changes in system design, or for which data are readily available in printed sources, such as soil surveys, or easily obtainable in the field.

The large differences in physicochemical properties among the three groups of organisms considered dictate that their removal be examined separately, since the factors influencing their fate are likely to differ. For example, when removal of fecal coliforms, a commonly used indicator of water contamination by STE, and viruses are compared (Figure 2-11), there are several instances where bacterial removal rates overestimate virus removal rates, while in other instances virus removal overestimates bacterial removal. Although seemingly small, the differences in removal rates can translate into large differences in bacterial or viral numbers because they are often present in STE at high concentrations. For example, a review by Lowe et al. (2007) found reports of 10^{3}-10^{8} fecal coliforms/100 ml of STE. A removal rate of 99% would still leave up to 10^{6}/100 ml. In comparison, the EPA's National Primary Drinking Water Standards call for a maximum contaminant level (MCL) goal for fecal coliforms of zero per 100 ml (U.S. EPA, 2008a). The use of fecal coliform removal as a predictor of viral fate is also questionable because viruses, through their protective protein coats and envelopes, are more resistant to environmental stresses and remain infective in the soil longer than bacteria (Canter and Knox, 1985).

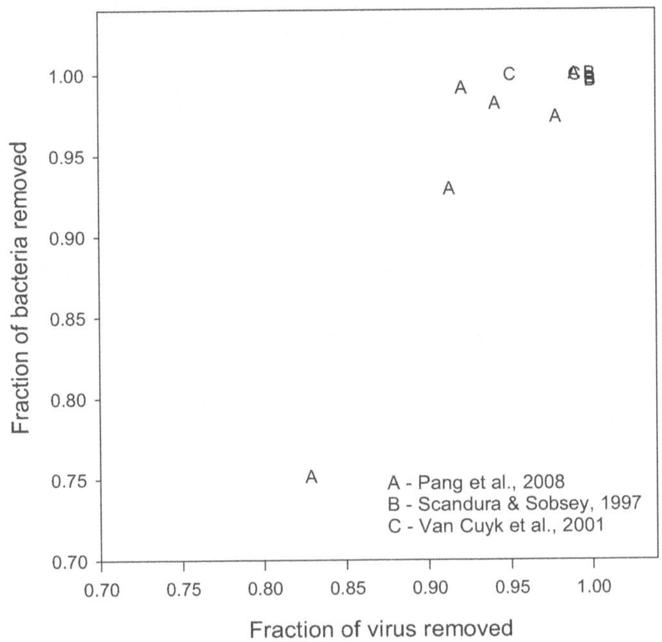

Figure 2-11. Relationship between Fraction of Bacteria Removed and Fraction of Virus Removed in Soil (dotted line indicates a 1:1 relationship).

2.5.2 Viruses

Viruses are extremely small (20-350 nm) (Bitton, 1980) infectious particles that cannot replicate outside of a host cell. They consist of single or double-stranded DNA or RNA within a protein shell, which in some cases is covered by an envelope of lipids or lipoproteins. Viruses replicate by attaching to and penetrating the host cell, taking over the host cell's own replication mechanism to multiply. Once the virus particles are mature, the host cell lyses to release the viruses. Viruses can infect all types of cells – plant, animal, and bacterial – but are usually specific to a certain host. Enteric viruses found in STE – which multiply in the intestinal tract of humans and animals and are excreted in large numbers – are of particular concern from a public health perspective. These include enteroviruses (poliovirus, echovirus, and cocksackievirus), Norwalk viruses, rotaviruses, reoviruses, caliciviruses, adenoviruses, and hepatitis A virus (Crites and Tchobanoglous, 1998). These viruses cause a variety of symptoms including severe diarrhea, respiratory illness, eye infections, and paralysis.

Enumeration of viruses often involves using a plaque-forming assay. The virus is mixed with a high density of the host cells and poured on a plate containing a solidified growth medium, such as nutrient agar. After incubation, zones of clearing, or plaques, are counted, with the assumption that each virus particle gave rise to one zone of clearing. The concentration of viruses is given in plaque forming units (pfu) per unit volume.

Because of the personal and public health risks posed by conducting studies using infective human viruses, studies of virus fate and transport in wastewater are often carried out using bacteriophages – bacterial viruses – as surrogates for human viruses Chu et al., 2003; (Powelson et al., 1990; Van Cuyk et al., 2004). For this purpose, a bacteriophage is chosen that has similar physical properties, such as size, shape, and isoelectric point (pI), to a human enteric virus. Proper matching is important because the physicochemical properties of viruses are thought to control their environmental fate. For example, differences in the size and conformation of the protein shell or envelope are thought to affect viral fate in soil, while the pI determines the net charge of the virus and thus its fate and transport. At pH values above the pI, the virus will have a net negative charge, while pH values below the pI result in a net positive charge. At low pH values, this may lead to an attraction to soil particles, which generally have a net negative charge, resulting in removal of the virus from the infiltrating wastewater.

Human viruses are not part of the normal fecal flora, but occur only in infected persons. This means that viruses are found intermittently in STE. It is estimated that the feces of an infected person may contain between 10^6 and 10^{10} viral particles/gram (Kowal, 1982; Feachem et al., 1983), resulting in their presence in STE at high levels. Hain and O'Brien (1979) sampled STE from three septic tanks and found numbers of enteroviruses capable of infecting HeLa cells (a human cell line) to range from 1.6 to 3.7 x10^3 pfu/L of STE. Feachem et al. (1980) reported numbers of enterovirus in STE to be 0 to 10^8 viruses/L of STE. Because the infective dose for viral infection can be as low as 1 organism (Bitton, 1980), and the concentration of virus in STE can be high, very high removal rates in the STU are necessary to minimize their transport to ground and surface water.

When viruses do penetrate the vadose zone below the STU and enter into groundwater, they have extended survival times and are capable of moving long distances in groundwater. For example, Vaughn et al. (1983), in a study of a shallow, sandy soil aquifer contaminated with a wastewater leachate plume, recovered human virus particles at a horizontal distance of 67 m (219.8 ft) and a vertical distance of 18 m (59 ft) from a leaching pool system. Scandura and Sobsey (1997) seeded septic systems with a bovine enterovirus and found rapid and extensive contamination of groundwater, with 908 pfu/L present 2 days after seeding at a distance of 35 m (114.8 ft). Schaub and Sorber (1977) found indigenous enteroviruses present 18 m (59 ft) below and 183 m (600.4 ft) away from a rapid infiltration site composed of silty sand and gravel. Viruses in surface water can also travel significant distances. In a study of marine viruses in Lake Geneva, Switzerland, high concentrations were found up to 3 km (1.9 mi) from the injection site (Goldscheider et al., 2007).

2.5.2.1 Removal/Inactivation of Viruses

Passage through the septic tank does not dramatically decrease virus populations (Canter and Knox, 1985; Van Cuyk et al., 2001). Feachem (1980) reports that septic tanks remove only 50% of virus from wastewater inputs. Thus septic systems rely on the STU to remove viruses before they reach groundwater or surface water. Many researchers have attempted to identify specific soil properties important to virus removal (Chu et al., 2003; Goyal and Gerba, 1979; Han et al., 2006), with most published studies limited to column studies and a few field investigations (Field et al., 2007; Scandura and Sobsey, 1997).

Because of the extremely small size of most viruses, there is thought to be limited removal through mechanical filtration. Hence, removal of viruses in the subsoil, when it takes place, is mainly attributed to adsorption, which depends on pH, OM content (Powelson et al., 1991), exchangeable Fe content (Gerba et al., 1975), soil texture (Gerba et al., 1975), presence of divalent cations (Moore et al., 1982), extent of water saturation (Lance and Gerba, 1984; Powelson and Gerba, 1994; Sobsey et al., 1980;), and the particular strain of virus (Goyal and Gerba, 1979). The multiplicity of factors controlling virus transport makes it difficult to pinpoint which are most important. Indeed, different factors may be important for different combinations of soil and virus.

2.5.2.2 Role of Soil Texture, Structure and Depth

Figures 2-12 and 2-13 summarize data from eight different soil column studies that examined virus removal as a function of soil texture and depth. These data show that neither texture nor depth of soil, which may be important for the removal of larger particles, correlate particularly well with virus removal. Where more than one value of removal rate is shown for a particular depth or texture, other properties may have varied within a particular experiment. These results are in agreement with the hypothesis that virus removal does not take place by filtration – at least not in laboratory-based column experiments, wherein the water content is generally strictly controlled. In field situations, however, texture and depth may play a role in virus removal by dictating whether saturated or unsaturated conditions prevail in the soil below the infiltrative surface. As stated previously, soil structure may also be important in virus removal by determining whether preferential flow paths may form, allowing percolating STE to bypass large volumes of soil, carrying virus particles with it. Preferential flow paths are associated with well-structured soils, such as those with high clay content. In a study using intact soil cores, Pang et al. (2008) found a viral reduction of only 0.1 to 2 orders of magnitude m^{-1} in a clayey soil in comparison with other, less structured, soils which removed 2 to 3 orders of magnitude m^{-1}. McLeod et al. (2001), using the same intact soil core method, found that large numbers of a *Salmonella* phage exited from a well-structured clay soil and from a sandy soil that was shown to foster finger flow.

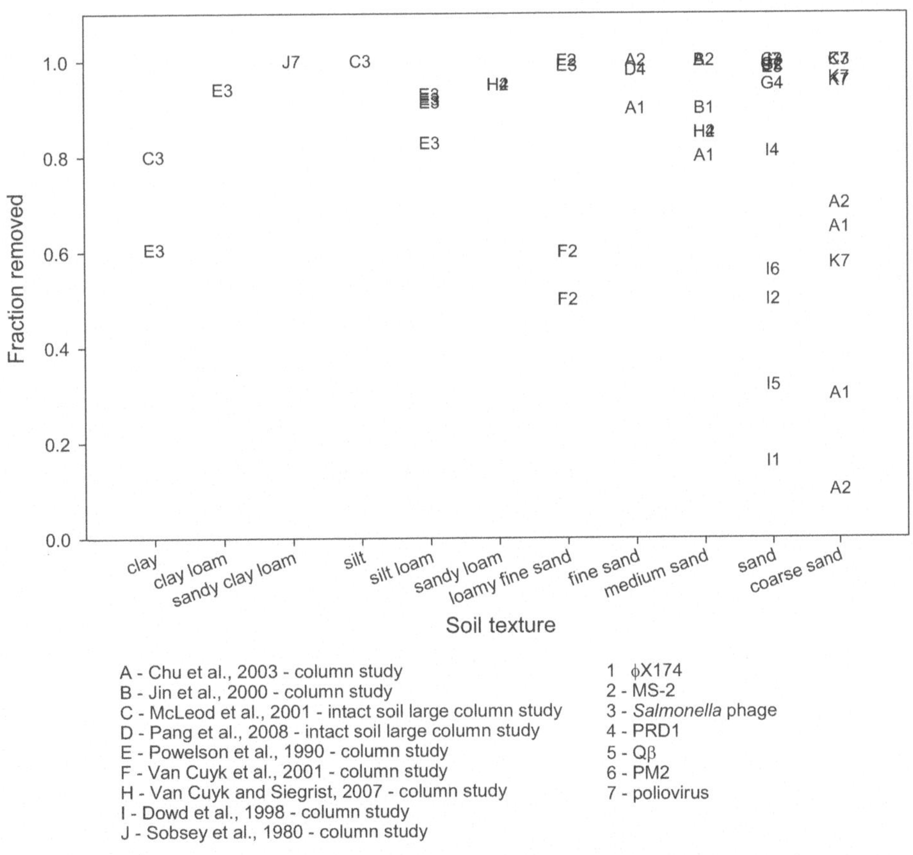

Figure 2-12. Relationship between Soil Texture and Fraction of Viruses Removed in Soil.
(dotted line indicates complete removal)

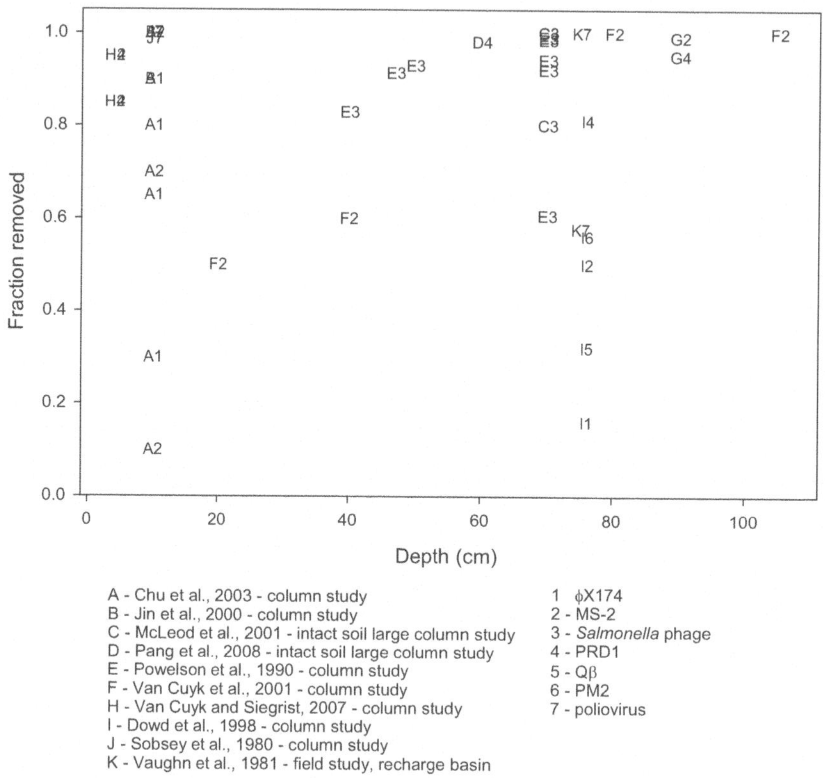

Figure 2-13. Relationship between Treatment Depth and Fraction of Viruses Removed in Soil. Dotted line indicates complete removal.

2.5.2.3 Role of the Unsaturated Zone

Many studies have shown that the presence of an unsaturated zone is important for virus removal (Powelson et al., 1990; Powelson and Gerba, 1994; Sobsey et al., 1980). This may be due to the establishment of a negative water potential gradient, which causes water to flow into finer soil pores, allowing for more contact of virus particles with soil particle surfaces (U.S. EPA, 2002a), or to partitioning of the virus to the air-water interface Powelson et al., 1990; (Trouwborst et al., 1974). Sobsey et al. (1980) found that sandy and organic soils showed poor viral adsorption for poliovirus and reovirus under saturated conditions, but removed 95% of viruses suspended in wastewater under unsaturated conditions. However, after subsequent addition of simulated rainfall, considerable quantities of virus were eluted. Conversely, columns containing clay soils under unsaturated flow conditions removed >99% of virus and did not release it under simulated rainfall conditions. Powelson et al. (1990) found little adsorption of bacteriophage MS-2 in a loamy fine sand under saturated conditions, but strong removal under unsaturated conditions. In addition, only 39% of the virus could be accounted for under unsaturated conditions (vs. 100% under saturated conditions) after desorbing the virus from the soil, indicating that virus inactivation took place. In a sandy alluvial soil, Powelson and Gerba (1994) found that virus removal rates were three times higher under unsaturated than saturated flow for the three types of virus studied – MS-2, PRD-1, and poliovirus.

As long as sufficient separation distance exits between the infiltrative surface and the groundwater, unsaturated conditions can be maintained by manipulating HLR and dosing schedules. Vaughn et al. (1981), in a study of groundwater recharge using tertiary treated effluent, found that high loading rates (75-100 cm/h; 2.5-3.3 ft/h) led to removal rates of only

58% for poliovirus, while a rate of 6 cm/h (0.2 ft/h) improved removal rates to >99% through 75 cm (2.5 ft) of soil.

2.5.2.4 Role of Soil pH

Batch studies – in which a virus suspension and soil are mixed in a flask under laboratory conditions – show that adsorption of the virus to the soil is more efficient when the pH of the soil is below the isoelectric point (pI) of the virus. Under these conditions the viral particle either has no charge or a net positive charge (Gerba et al., 1975). According to Gerba et al. (1975), the pI for most enteroviruses is below 5.0. In batch studies with nine different soils conducted by Goyal and Gerba (1979), pH was found to be the only significant factor in controlling adsorption for five different viruses, and a negative correlation, although not significant at $P < 0.05$, was observed with pH for adsorption of four other viruses. Soil below pH 5.0 was found to be the best adsorbent. In the case of soils receiving STE, which has a near neutral or slightly basic pH (Canter and Knox, 1985), the saturated soil at the infiltrative surface tends to have a pH near neutral, which is not optimal for virus adsorption. However, the unsaturated zone below the biomat may possess an acidic pH. This was illustrated in studies using STU mesocosms where Potts et al. (2004) and Patenaude et al. (2008) found the pH of water draining from intermittently aerated, unsaturated lysimeters to be approximately 4.5 even though STE inputs had a neutral pH, probably due to acidifying processes such as nitrification.

Although Goyal and Gerba (1979) propose that virus adsorption is better at a pH below 5.0, analysis of results from nine different published studies on virus removal suggests that this is not always the case (Figure 2-14). Notably, there are few studies using soil with a pH of <5.0. Furthermore, it appears that at pH > 7.0 virus removal decreases, which is in agreement with Canter and Knox (1985), who indicate that virus removal is rapid and effective at pH < 7.4. This indicates that there may be an optimal ratio between the number of sites in the soil available for adsorption and the ideal pH for virus adsorption based on the isoelectric point (pI). A soil's generally net negative charge is primarily associated with the colloidal fraction of soil, which consists of clay-sized particles and humus. Colloids are considered highly reactive because, although they are very small (diameter <2 μm), they have a larger specific surface area per unit mass for surface reactions, such as virus adsorption, to take place relative to larger soil particles, such as sand.

Humus, which has a high net negative charge density under neutral or alkaline conditions, probably does not play a role in virus adsorption in the subsoil, since it is naturally found in low concentration in the lower regions of the soil profile. In contrast, clay minerals can be present throughout the soil profile, including soil below the infiltrative surface. The permanent, or pH-independent, negative charge on clay minerals comes from isomorphous replacement of ions (e.g., Al^{3+} with Mg^{2+}). The variable, or pH-dependent, negative charge of clay minerals is associated with OH^- groups on the edges of clay plates and charges associated with Al- and Fe-hydroxides (Stevenson, 1994; Brady and Weil, 2002). At low or moderately acid pH values, clay minerals lose their net negative charge because of protonation of reactive groups (e.g., a hydrogen ion, H^+, is attached to the negatively charged sites), lowering their ability to interact with virus particles. The term "clay" can refer to clay-sized particles or to clay minerals. Studies of virus adsorption often refer to specific clay minerals, but others do not distinguish between the two. This makes interpretation of the results difficult. For instance, Scandura and Sobsey (1997) state that soils containing a minimum of 15% clay are good for virus removal. However, these authors did not distinguish between clay-sized particles and clay minerals, but described the soil used in their study as being in the Norfolk series and kaolinitic. If so, the observed virus removal may be due to the high density of negative charges on kaolinite colloids.

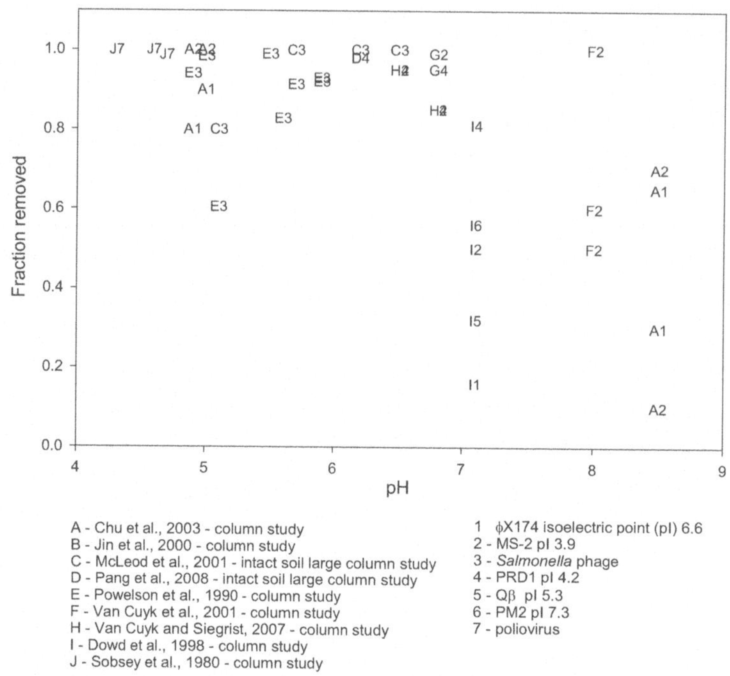

Figure 2-14. Relationship between pH and Fraction of Viruses Removed in Soil.
(dotted line indicates complete removal)

2.5.2.5 Role of Organic Matter

Because inputs of plant detritus generally enter the soil at or near the soil surface, STU soils, which are considerably deeper in the soil profile, typically have low levels of soil OM. In contrast, STE has high concentrations of dissolved and particulate organic material, usually measured as 5-day biochemical oxygen demand (BOD_5) and total suspended solids (TSS). Organic material is primarily removed in the septic tank and at the infiltrative surface (U.S. EPA, 2002a). However, a fraction of the colloidal and soluble organic material in STE penetrates into the unsaturated zone, where it can compete with viruses for adsorption on clay particles (Crites, 1985). Pieper et al. (1997) found that three times more bacteriophage PRD-1 moved through an aquifer contaminated with sewage than through an uncontaminated zone. Similarly, Powelson et al. (1991) concluded that virus removal was enhanced in columns containing lower levels of dissolved and colloidal OM.

2.5.2.6 Miscellaneous Factors

Other properties found to affect virus adsorption include the ionic strength of the infiltrating water and hydrophobic partitioning. Adsorption of viruses may occur even when soils and viruses are both carrying a negative charge. In waters of near neutral pH with high ionic strength (such as STE), divalent cations such as Ca^{2+} and Mg^{2+} can coat the viral surfaces, giving them a positive charge which allows them to attach to soil particles (Canter and Knox, 1985). Han et al. (2006) found that in columns filled with chemically heterogeneous (hydrophobic and hydrophilic) glass beads, virus retention increased as the ionic strength of the medium increased. Hydrophobic partitioning, which involves interaction between hydrophobic regions of virus particles and those found in soil OM, also has been suggested as a mechanism for virus removal (Bales et al., 1991).

Age of the STU may also affect virus adsorption. Van Cuyk and Siegrist (2007), in a mini (4 cm) column experiment aimed to model effects of the changing infiltrative surface on virus removal, found that as the system aged and a biofilm began to form, removal of MS-2 and PRD-1 generally improved.

Faulkner et al. (2003) notes that septic system effluents can be expected to contain significant amounts of household surfactants. In this context, Thompson and Yates (1998) demonstrated that addition of a surfactant can enhance MS-2 survivability by the concurrent increase in air-water interfacial tension. More recently Chattopadhyay et al. (2002) found that addition of anionic surfactants tends to increase viral inactivation, and that the effects of surfactants on sorption varies depending upon the hydrophobicity of viral surfaces.

2.5.3 Bacteria

Bacteria are single-celled microorganisms with diameters ranging from 0.3 to 2.0 μm. Unlike eukaryotic cells that contain membrane-bound organelles, bacteria do not contain a membrane-bound nucleus, and are considered prokaryotes. Most species replicate by binary fission. There are about 10^{13} bacteria in the human digestive tract, with members from 500 to 1000 species present (Sears, 2005). In the case of individuals infected with a pathogenic bacterial species, large numbers may be shed in the feces. For example, *Salmonella* can be present in STE at concentrations of 10^7 cells/100 ml (Feachem et al., 1980). Other human-pathogenic bacterial species can be present in wastewater and STE at levels ranging from 10^3 to 10^8 (Table 2-4). These bacteria can cause symptoms ranging from severe diarrhea to acute respiratory illness. In the case of immune-compromised individuals or the very young or elderly, death may result from infection. The infectious dose for bacterial species covers a wide range (Table 2-4), but infections can be caused by as few as 10 cells. As with viruses, this low infective dose coupled with the high numbers that may be present in STE from an infected individual's household, leads to a requirement for extremely high removal rates exceeding three orders of magnitude.

While 50-90% of the pathogenic bacterial load of raw wastewater may be removed in the septic tank (Feachem et al., 1980), removal in the STU is a function of predation (Crites, 1985), exposure to adverse conditions such as low temperatures (Crites, 1985), inability to compete with indigenous bacteria (Yates and Gerba, 1998), mechanical filtration (Canter and Knox, 1985), adsorption (Canter and Knox, 1985), sedimentation (Sobsey and Shields, 1987), and flow rate (Ausland et al., 2002). Much of our understanding of the fate of fecal bacteria in STU soils stems from studies using fecal coliforms, *E. coli*, and fecal enterococci. These organisms are used as indicators of human fecal contamination in ground and surface waters, and some members of these groups are human pathogens. Because bacterial species vary widely in their ecology and physicochemical properties, the fate of these organisms is likely not representative of all enteropathogenic bacteria found in STE.

2.5.3.1 Role of Filtration

Unlike reduction of viruses, mechanical filtration plays an important role in the removal of bacteria from STE. This is due to the comparatively greater size of bacteria, which are approximately 100 times larger than viruses. Filtration takes place at the biomat (Gerba et al., 1975) and/or in the underlying unsaturated zone. Although various studies point to the importance of the biomat in bacterial removal (e.g., Van Cuyk and Siegrist, 2001), Potts et al. (2004) in their study of aerated STU mesocosms showed that a conventional biomat may not be necessary for high removal rates of fecal coliforms if it is replaced with a zone of active aeration. Miles et al. (2007) also noted the importance of a zone of aeration around infiltrative surfaces to provide for coliform removal in pressure-dosed systems. Bacteria that do infiltrate through the biomat can be removed by filtration in the underlying soil if unsaturated conditions prevail (Stevik et al., 2004). Under these conditions, water moves into smaller pores at a lower velocity, which may result in higher removal rates than those found in saturated conditions, where water flows at a higher velocity and bypasses small pores.

2.5.3.2 Role of Texture and Depth

Soil texture plays a part in mechanical filtration of bacteria by dictating soil pore size distribution. For example, Ausland et al. (2002) found that fecal coliform removal was more than three orders of magnitude greater in unsorted fine sand than in 2 to 4 mm aggregates. However, analysis of data from 12 studies, four of which are field-based, suggests that, although coarser textured soils are much less efficient at removing bacteria than finer textured soils, the depth of the treatment soil is more important (Figures 2-15 and 2-16). Typically, a minimum of 0.5 m (1.6 ft) of soil is needed for removal close to 100% (Figure 2-16). Our analysis of the reviewed data is in agreement with the findings of Hagedorn (1984), who concluded in a review article that 0.3-0.9 m (12-36 in.) of unsaturated soil were necessary for acceptable bacterial removal rates, although Karathanasis et al. (2006) (Figure 2-16) saw removal rates for fecal coliform bacteria of only 64% in 60 cm (2 ft) of a clay loam soil. Similar conclusions were drawn by Beal et al. (2005) and Gill et al. (2007), who suggest a slightly greater depth of 0.6-0.9 m (24-36 in.) of unsaturated soil is necessary. Karathanasis et al. (2006) in a study that evaluated bacterial removal at soil depths of 0.3, 0.45, and 0.6 m (12, 18, and 24 in.) using undisturbed soil monoliths of ten different soil types found that 0.6 m (24 in.) of soil provided the best and most consistent performance regardless of soil texture. Karathanasis et al. (2006) hypothesized that soil macroporosity, which results from more defined soil structure, is more common in some finer textured soils. Higher macroporosity would lead to preferential flow through the larger pores and, therefore, less removal of bacteria through mechanical filtration.

Conversely, in a study of intermittently aerated leachfield mesocosms containing 7.5, 15, or 30 cm of silica sand (diameter 0.71-0.21), URI found the removal of fecal coliform bacteria over time to be unaffected by depth (Amador et al., 2008). Average removal rates were 82.4%, 92.3%, and 85.7% for the three depths, respectively. A previously published study (Atoyan et al., 2007), using 30 cm of native soil (92% sand, 8% silt), found removal rates ranging from 99.94% to 100%. These results suggest that at depths ≤ 30 cm, depth may not be important for pathogen removal. Higher removal rates found in the Amador et al. (2008) soil mesocosm study at 30 cm may be due to differences in texture; however, they may also be due to differences in the age of the systems. The soil mesocosm study used lysimeters that had been in operation for approximately three years prior to the start of the experiment, while the sand mesocosm study examined lysimeters for the first 11 months of operation. Although the presence of a biomat was not observed in either study (intermittent aeration serving to prevent its formation), longer running times may serve to allow for development of biotic and abiotic conditions favorable for bacterial removal.

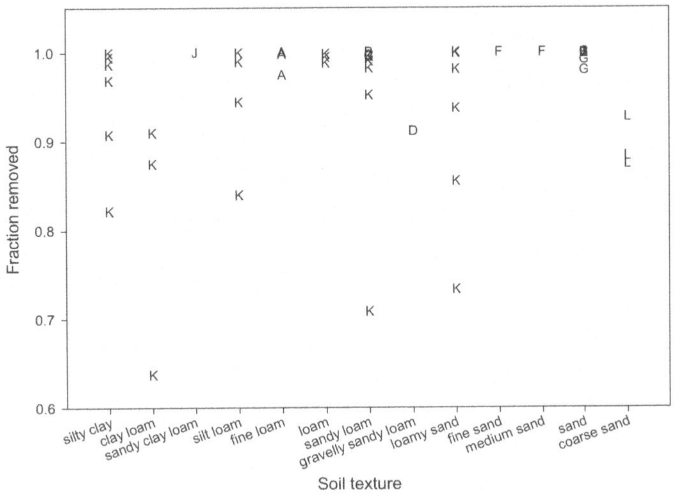

A - Duncan et al., 1994 - column study, undisturbed soil
B - Field et al., 2007 - field study of conventional systems and seepage pits
C - Scandura & Sobsey, 1997 - field study of conventional systsms
D - Harrison et al., 2000 - field study
E - Atoyan et al., 2007 - column study
F - Stevik et al., 1999 - column study
G - Potts et al., 2004 - column study
H - Van Cuyk et al., 2001 - column study
I - Ausland et al., 2002 - field study of filters
J - Sobsey et al., 1980 - column study
K - Karathanasis et al., 2006 - column study, undisturbed soil
L - Amador et al., in press - column study

Figure 2-15. Relationship between Soil Texture and Fraction of Fecal Coliforms and *E. coli* Removed in Soil.
(dotted line indicates complete removal)

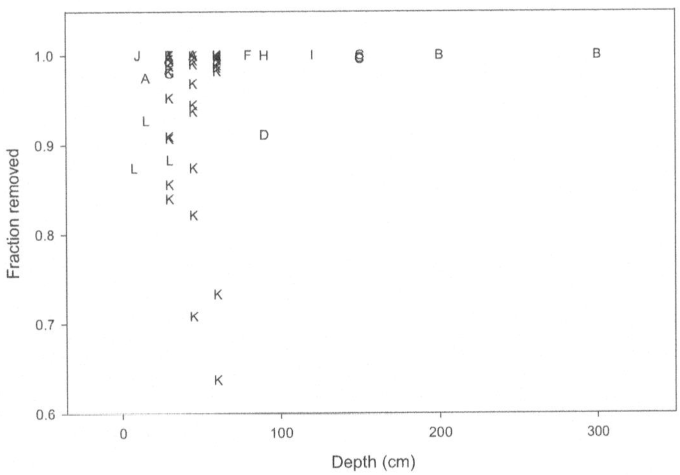

A - Duncan et al., 1994 - column study, undisturbed soil
B - Field et al., 2007 - field study of conventional systems and seepage pits
C - Scandura & Sobsey, 1997 - field study of conventional systsms
D - Harrison et al., 2000 - field study
E - Atoyan et al., 2007 - column study
F - Stevik et al., 1999 - column study
G - Potts et al., 2004 - column study
H - Van Cuyk et al., 2001 - column study
I - Ausland et al., 2002 - field study of filters
J - Sobsey et al., 1980 - column study
K - Karathanasis et al., 2006 - column study, undisturbed soil
L - Amador et al., in press - column study

Figure 2-16. Relationship between Treatment Depth and Fraction of Fecal Coliform Bacteria and *E. coli* Removed in Soil.
(dotted line indicates complete removal)

2.5.3.3 Extent of Soil Saturation

As is the case with viruses, the presence of an unsaturated zone, which results in a prolonged STE retention time and maximum contact of STE with the soil, is thought to improve bacterial removal rates (Beal et al., 2005). This is due to forcing water into smaller soil pores, where bacteria will be either filtered out or more effectively adsorbed to soil. In addition, unsaturated flow is less susceptible to forming preferential flow paths. Higher aeration, which is associated with coarser-textured unsaturated soils, may also support higher numbers of protozoa and nematodes – both of which graze on bacteria. This may lead to higher removal rates of pathogenic bacteria (Amador et al., 2006). However, Harrison et al. (2000) in a field study of septic systems located in coarse-textured glacial outwash, found a removal rate for fecal coliform bacteria of only 91% through 1 m (3.3 ft), resulting in 10^6-10^7 bacteria/L.

2.5.3.4 Hydraulic Loading Rate

Individually, a number of studies using variable HLRs (Ausland et al., 2002; Potts et al., 2004; Stevik et al., 1999) suggest that an increase in HLR corresponds with a decrease in bacterial removal. However, when these data are analyzed together, the effects of HLR are not as clear (Figure 2-17). For example, Van Cuyk et al. (2001) found identical removal rates for HLRs of 5 and 8.4 cm/day (2 and 3.3 in./day). Analysis of results from eight laboratory studies with HLRs ranging from 1.5 to 12 cm/day (0.6 to 4.7 in./day) (Figure 2-17) shows that there is considerable variability in removal rates at loading rates <4 cm/day. In general, it appears that bacterial removal is not particularly sensitive to HLR. This is counterintuitive, since higher rates reduce retention time, increase soil moisture, and increase the probability of formation of preferential flow paths. This is thought to result in lower removal of bacteria as there is less time for removal processes to occur in the STU. Other factors, such as oxygen availability, and system age, may be more important than HLR in controlling bacterial removal.

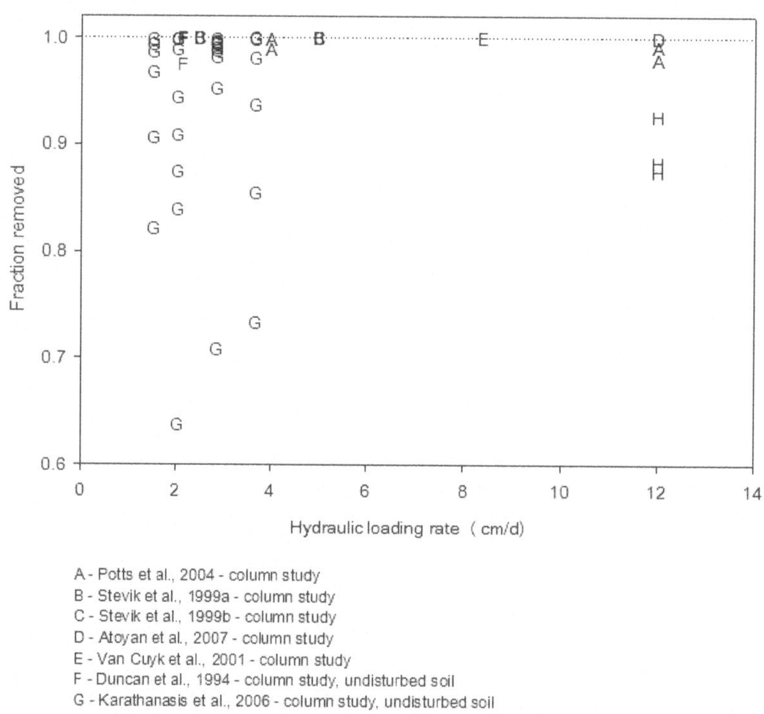

Figure 2-17. Relationship between Hydraulic Loading Rate and Fraction of Fecal Coliform Bacteria and *E. coli* Removed in Soil. (dotted line indicates complete removal)

2.5.4 Protozoa

Protozoa are unicellular, eukaryotic organisms – they contain membrane-bound organelles. Most species of protozoa are not harmful, and may actually have a beneficial role in wastewater microbiology by preying on pathogenic bacteria (Horan, 2003). Enteropathogenic protozoa, those capable of causing disease in the intestinal tract, can have a much more complex reproductive cycle than bacteria, including the formation of survival forms, such as cysts or oocysts. These cysts and oocysts are then excreted in large numbers from infected individuals. Cysts are very resistant to inactivation and can survive for long periods in the environment. Human-pathogenic protozoa, such as *Cryptosporidium parvum, Entameoba histolytica,* and *Giardia lamblia*, are not a normal component of the STE microflora and are present only when an individual is infected. When present, levels of cysts in STE range from 10^1 to 10^5/100 ml (Feachem et al., 1980). *Giardia* cysts range in size from 8 to 14 µm in length, *Entameoba* cysts are 10 to 15 µm in diameter, and *Cryptosporidium* oocysts are usually 4 to 6 µm in diameter (Crites and Tchobanoglous, 1998). As such, they are about 10 times the size of bacteria and mechanical filtration is likely to be the primary means of removal from STE. Microsporidia, such as *Encephalitozoon sp.,* are also of concern. These are less likely to be removed through mechanical filtration, however, because of the extremely small size of their spores (1-5 µm). The infective dose for these protozoan species is often very low (Table 2-4) – theoretically, ingestion of one cell can cause disease – so high removal rates are necessary. Only a few studies are available on the removal of protozoan cysts in STU.

2.5.4.1 Role of Texture and Depth

According to Feachem et al. (1980), there may be no removal of protozoan cysts or oocysts (the forms that are found in the environment) in the septic tank. Because the primary removal mechanism for protozoan cysts is thought to be mechanical filtration, the soil texture, which dictates pore sizes, should play an important part in removal. For instance, Darnault et al. (2003) in a study of *C. parvum* removal found that 18 times more oocysts were present in the effluent from 18 cm thick profile of coarse-textured sand than from fine-textured sand. Less removal was seen in an undisturbed fine soil. This was probably due to preferential flow through macropores and wormholes. The influence of macropores was also highlighted by Bradford et al. (2006), who found few *G. lamblia* cysts transported through different grain-sized sands when suspended in a buffer solution. When manure was added to the suspension, the effluent concentration of cysts increased, due to filling of smaller straining sites. Analysis of data from three column studies (Figure 2-18) shows that removal of *C. parvum* oocysts is a function of soil texture. However, removal is less than 99% in all cases, sometimes greatly so, which may be due to differences in the depths of soil columns tested. The studies reviewed in the analysis were all column studies with media depths ≤18 cm, where depth does not seem to correlate well with removal (Figure 2-19). However, greater removal rates might be found with soil depths more closely resembling those found in STUs.

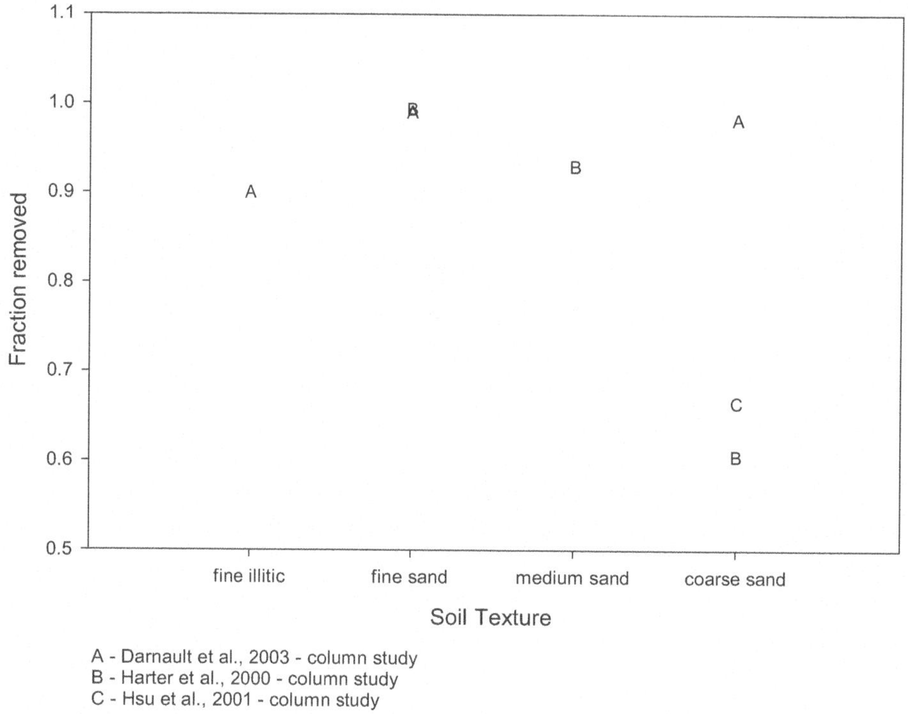

Figure 2-18. Relationship between Soil Texture and Fraction of *C. parvum* oocysts Removed in Soil. (dotted line indicates complete removal)

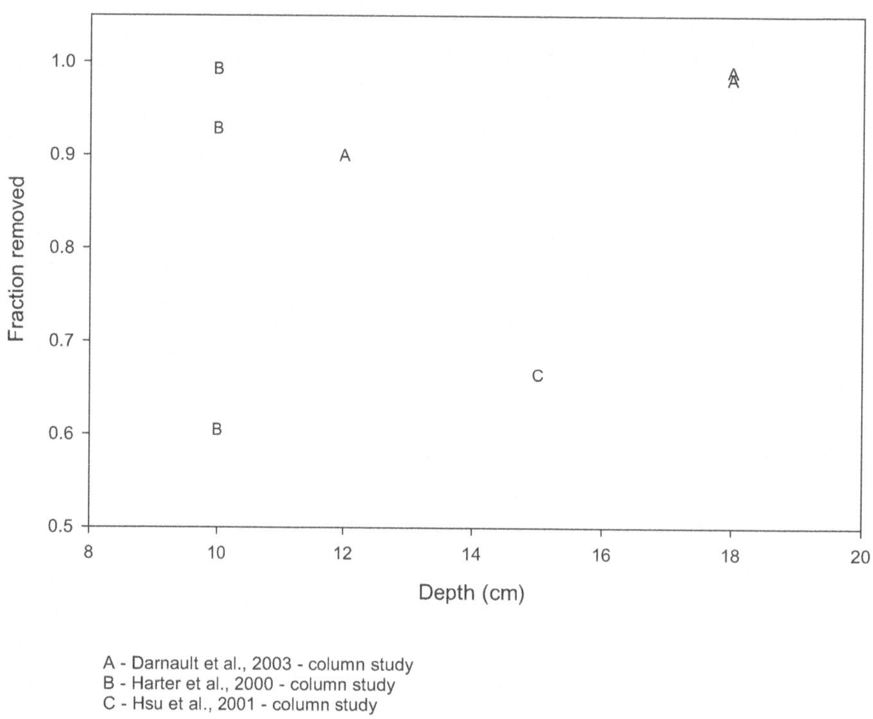

Figure 2-19. Relationship between Treatment Depth and Fraction of *C. parvum* oocysts Removed in Soil. (dotted line indicates complete removal)

2.5.4.2 Additional Factors

Bradford et al. (2006) suggests that, for smaller sized protozoa, chemical properties of both the organism and the infiltrating solution also play a part in dictating removal rates, although few studies have examined these variables. Hsu et al. (2001) examined the effects of ionic strength and pH on the removal of *C. parvum* and *G. lamblia* cysts. Higher electrolyte concentrations improved the collision efficiencies of both species leading to greater removal, while lower pH improved the removal of *Cryptosporidium*.

2.5.5 Conclusions

Based on examination of the literature, the following factors primarily control the fate and transport of viruses, bacteria, and protozoa in soil (Table 2-5):

Viruses: Neither soil texture nor treatment depth appear to be important controls on viral fate. In contrast, the interplay of virus isoelectric point, pH and clay mineralogy appears to be an important factor in determining virus removal, as is the level of dissolved OM in STE and the presence of unsaturated conditions below the infiltrative surface.

Bacteria: Removal of bacterial pathogens takes place primarily by mechanical filtration, a process that is governed by soil texture, treatment depth, and the presence of unsaturated conditions below the infiltrative surface. HLR does not appear to have a consistent effect on bacterial removal.

Protozoa: Mechanical filtration appears to be the main process for removal of protozoan cysts and oocysts from STE.

Table 2-5. Summary of System and Organism Properties and Their Effects on Pathogen Removal Rate.

Pathogen	Property	Effect on Removal Rate
Viruses	pH	✓
	Clay mineralogy	✓
	Soil texture	–
	Viral pI	✓
	Treatment depth	–
	Dissolved OM	✓
	Unsaturated conditions	✓
Bacteria	Soil texture	✓
	Treatment depth	✓
	Hydraulic loading rate	–
	Unsaturated conditions	✓
Protozoa	Soil texture	✓
	Treatment depth	✓

✓ = important effect
– = little effect

Optimizing removal of pathogenic organisms in conventional soil-based systems is a challenge because different factors control the fate of viruses, bacteria, and protozoa in STE. In principle, the use of soils with a sufficient amount of appropriate clay minerals would increase the probability of optimizing removal of viruses, bacteria and protozoa. However, soils with these properties are not evenly distributed in space, either requiring the use of engineered soils using imported clays minerals, or foregoing the benefits of virus removal in areas that lack appropriate clay minerals. In addition, the presence of high levels of clay minerals restricts infiltration, which would require an increase in the system's infiltrative surface area. Alterations in these design parameters are also likely to affect biogeochemical processes involved in removal of nutrients, such as P and N, which rely on a particular set of environmental conditions and/or sequence of events to take place.

The majority of available peer-reviewed studies have been conducted at laboratory or pilot scales. Field-scale evaluations of pathogen removal are scant, and they focus on a particular type of organism, with little or no consideration of other functions performed by the STU. Future research should integrate field evaluation of the role of system design and soil properties in determining removal of different types of enteropathogenic organisms within the context of other STU functions, such as infiltration and nutrient removal. These data can then be used to develop predictive mathematical models that describe the various functions of the STU in terms of design parameters and available information on soil properties. Simplified versions of these models can be made available to designers to help improve system performance.

2.6 Organic Pollutants

2.6.1 Background

Organic wastewater contaminants – often referred to as emerging contaminants, pharmaceuticals and personal care products, or organic micropollutants – are being detected in surface water and groundwater influenced by OWTS (Drewes et al., 2003; Hinkle et al., 2005; Zimmerman, 2004). This suggests that some current wastewater treatment systems are unable to eliminate these compounds from the effluent. Organic wastewater contaminants (OWC) include pharmaceuticals, detergents, surfactants, antimicrobial additives, toothpaste, and other personal care and cleaning products.

The presence of OWCs in the aquatic environment is receiving increased public attention. Discover Magazine named the Toxic Substances Hydrology Program's study "National Reconnaissance of Pharmaceuticals, Hormones, and Other Organic Wastewater Contaminants in Streams" conducted by Kolpin et al. (2001) as one of the top 100 science stories of 2002 (Kolpin et al., 2001; Discover, 2002). In March of 2008, Popular Science magazine ran an article entitled "Tainted Tap Water" which focused on an Associated Press story claiming it had found "traces of dozens of pharmaceuticals in the drinking water of an estimated 41 million Americans" (Dyer, 2008). These two articles are just a small sample of the heightened public attention this topic is receiving.

Within the Small Flows Program at the CSM Environmental Science and Engineering Division, Conn (2008) investigated the occurrence of OWCs in OWTS effluents as well as OWC concentrations with depth at the Mines Park Test Site. Preliminary indications suggest that triclosan was reduced to insignificant concentrations within 1.2 meters of infiltration in a sandy loam soil while nonylphenol was observed in the microgram per liter range after infiltrating through 2.5 meters of the STU.

Several government agencies have invested in research efforts and are considering regulation of OWCs, demonstrating the perceived importance of these compounds as environmental contaminants. The U.S. Geological Survey (USGS) and the U.S. Environmental Protection Agency (U.S. EPA) are currently involved in many studies examining OWCs in wastewater. The U.S. EPA's informational website http://www.epa.gov/ppcp includes a list of the current research efforts.

The U.S. EPA moved toward applying water quality criteria to select OWCs with the release of a white paper report on June 3, 2008 (Aguirre et al., 2008). The water quality criteria will be designed to protect aquatic life from a list of contaminants of emerging concern largely populated with OWCs. The listed contaminants have not yet been made public. This report uses 17β-estradiol as an example for how to quantify effects on ecological systems. Some of these contaminants may demonstrate low acute toxicity, but cause significant reproductive effects in some organisms and thus require special attention. Because of this, the chronic toxicity tests in the 1985 guidelines for water quality criteria are not sufficiently comprehensive for OWCs. The white paper thus recognizes that regulations may need to be considered for some contaminants that do not have sufficient toxicity data to meet minimum guideline standards.

Current USGS projects are posted on the website http://toxics.usgs.gov/regional/emc/. The U.S. EPA, USGS, and USDA have collaborated to form the Interagency Subcommittee for Pharmaceuticals in the Environment (PIE) which seeks to create a federal plan to research the issue of pharmaceuticals in drinking water. The report, originally expected to be released in the spring of 2008, has not been made publically available as of February 2009. An EPA informational website (http://www.epa.gov/waterscience/ppcp/basic.html) summarizes the topic and has a section addressing regulatory action. As of February 5, 2009 no specific contaminants are identified as requiring regulation. Statements have been made by various PIE personnel recognizing the ubiquitous appearance of pharmaceutical and personal care products in the environments. Modes of detections are described along with assurance of ongoing research and the need for regulation is expressed but no details are given (Hirsch, 2008; Larsen, 2008).

Recent newspaper articles criticizing PIE's inability to make the report available demonstrate the ongoing public interest on the topic. The article, "Water task force misses target: Still no plan to deal with drug contamination," in the Seattle Post-Intelligencer, is an example of one of these articles (Mendoza, 2008).

While OWCs are receiving a lot of interest from the U.S. EPA, regulations and MCLs have yet to be set for most of these compounds. Nonylphenol is a notable exception. The aquatic life criteria for 4-nonylphenol is presented in Section 2.6.4. As more information about the appearance and effects of OWCs in the environment becomes available, it is likely that more of these compounds will be assigned similar regulations.

A thorough review of literature focusing mainly on the fate and transport of these compounds is summarized in this report. This review resulted in a selection of five out of hundreds of OWCs deemed most important and representative. The five chosen compounds are triclosan, nonylphenol, 1,4-dichlorobenzen, 17β-estradiol, and diclofenac. The information available for these five compounds is summarized in Appendix A.

2.6.2 Methods

Studies in peer-reviewed literature have focused on the occurrence and concentrations of selected compounds in streams, lakes, and groundwater impacted by wastewater. Most of the existing work focuses on scenarios where the OWC source to the environment was wastewater treatment plant effluents. For many compounds, the concentrations in the environment (streams, lakes, and rivers) are in the nanogram per liter (parts per trillion [ppt]) range (Kolpin et al., 2001). In septic tank and wastewater treatment plant effluent, some compounds are commonly found in the microgram per liter (parts per billion [ppb]) range (Conn et al., 2006). In general, compounds from personal care products are found in concentrations up to three orders of magnitude higher than pharmaceuticals (Conn et al., 2006; Daughton, 2004; Heberer, 2002; Jones et al., 2002; Kanda et al., 2003; Kinney et al., 2006; Kolpin et al., 2001; Moldovan, 2006). The aqueous concentrations in the environment may be low for OWCs but the nature of these organic molecules and the low concentrations at which some have been shown to cause toxic effects is reason for concern (Larson et al., 2004; Lienert et al., 2007, McLeese et al., 1981; Oaks et al., 2002). Because these molecules are organic, they may sorb to OM in soil and solids in wastewater, and hydrophobic/lipophilic qualities cause them to concentrate in the fats and lipids of organisms. The degree of this partitioning can be quantified using a bioconcentration factor which is the concentration of compound found in the organism divided by the concentration in the organism's habitat waters. Bioconcentration factors can be on the order of 1000 and higher for some organisms for many OWCs (Balmer et al., 2004; Coogan et al., 2007).

Organic wastewater contaminants can have a wide array of ecotoxicological effects and negative impacts. Endocrine disrupting compounds are a subclass of OWCs that negatively influence reproductive development and hormone production and response in some organisms. This has been shown to cause "gender-bender" effects in some species of fish and amphibians causing populations to become predominantly female or male organisms to exhibit female qualities (Brown et al., 2003). Some OWCs affect other internal systems such as the renal system which can lead to kidney failure in some organisms (Oaks et al., 2002). Other compounds have been shown to be carcinogenic and/or mutagenic (Robertson, 1994). Antimicrobial compounds may contribute to the evolution of resistant strains of bacteria (Glaser, 2004; Singh, 2007; Suarez et al., 2007).

Where OWTS sources are concerned, these compounds may be released into the STU from the effluent of the septic tank or other engineered treatment units. Based on the mechanisms for removal of these compounds from infiltrating water (aerobic biodegradation, sorption to soil particles), a well designed STU has the potential to reduce concentrations to negligible amounts for most OWCs, although the responsible microbiological, chemical, and physical mechanisms are not yet well understood. The fate and transport of the OWCs most likely to cause environmental and human health concerns in STUs needs to be characterized. The goals of this review are thus:

- Identify the five OWCs that are most likely to be regulated in the future.
- Use data and information from previous studies to best characterize relevant fate and transport characteristics in soil and groundwater for each contaminant.
- Describe and summarize how the existing knowledge is applicable and relates to transport, transformation, and treatment in STUs below OWTSs.
- Based on the existing information in the literature, hypothesize which STU design characteristics may be most effective in removal of each contaminant and in which STU zone treatment occurs.
- Identify gaps in the current knowledge and understanding of the processes relevant to aqueous transport of contaminants in STUs and propose experiments that may aid in filling these gaps.

To select the five OWCs for this study, many compounds were considered. Hundreds of OWCs have been identified in aquatic environments influenced by wastewater treatment system effluent. The compounds included in this list do not necessarily reflect the relative toxicity of compounds or the relative degree of occurrence. Some of these compounds, such as caffeine, are recalcitrant, but non-toxic and are studied simply as markers for wastewater in aquatic environments or the saturated subsurface (Verstraeten et al., 2005).

Priority OWCs were selected based on an assessment of the compounds that are most likely to have environmental implications and most likely to be regulated. To be considered a priority OWC, a chemical must meet four major criteria points:

- First, occurrence in onsite wastewater effluent must be ubiquitous; concentrations must be present in a large portion of onsite systems that utilize an STU.
- Second, mean concentrations of priority OWCs in onsite system effluent must be high enough to likely cause impacts in groundwater and surface waters. For most contaminants we have chosen the microgram per liter range (i.e., ppb). For hormones and some pharmaceuticals the nanogram per liter range (i.e., ppt) is significant.
- Third, priority OWCs must pose an ecotoxicological or human-health risk. Probable negative effects must be well documented in the literature.
- Finally, priority OWCs or harmful degradants (i.e., daughter products) must persist in the subsurface environment. Any compound that is likely to be transformed to harmless byproducts in the soil is less likely to reach stream or groundwater systems.

All OWCs found in the literature review were considered. For a complete list of sources see the biliography at the end of the report. Most contaminants were removed from the list because of evidence of no ecotoxicological effects or little to no appearance in OWTS. For this study, because of time limitations, complete analysis of selected compounds is not feasible. This constraint requires that a foundation of information about general fate and transport characteristics be available in the literature. Organic wastewater contaminants that have significant ecotoxicological effects and common appearance in wastewater are also more likely to have fate and transport study results. The remaining contaminants were rated qualitatively based on the criteria points. Table 2-6 summarizes this qualitative comparison for selected OWCs. This qualitative comparison considers information from peer-reviewed literature unique to each compound. Toxicity and risk, persistence in the environment, and appearance in onsite wastewater were evaluated based on studies available for the compound and qualitative evaluations were made for each criterion.

Toxicity/Risk. If toxic effects were shown to besevere, a high risk was assigned. If possible toxic effects have been studied but not found and concentrations of the molecule were high, very low was assigned but the contaminant remained on the list. If environmental ecotoxicological effects have been shown to be mild but bioconcentration was shown to be high, a moderate toxicity/risk was assigned.

Persistence. A contaminant that degraded completely in OWTS before reaching the STU was assigned a low persistence rating. A very high rating was assigned to contaminants that persist in the environment indefinitely. If nearly complete removal of the contaminant occured, "low" was assigned. If the contaminant is observed in the environment outside of OWTS, a minimum rating of moderate was assigned. If aqueous phase concentrations have been observed in surface waters or groundwater, persistence was considered to be high.

Appearance. If a compound infrequently appears in onsite wastewater, it is assigned low appearance. A compound that appears in low concentrations (less than 1 ng/L for most compounds) also is assigned low appearance. For each contaminant, the concentration considered significant depends on concentrations shown to have ecotoxicological effects. Significant concentrations may be much lower than toxic concentrations where accumulation or bioconcentration is relevant. Any concentration above 10 μg/L in septic tank/onsite system effluent or the environment is considered significant because of the continuous release of the compound from onsite systems. If a compound appears in over half of studied onsite wastewater that has been analyzed for the compounds in significant concentrations, moderate is assigned. A compound that appears in nearly all onsite wastewater in significant concentrations is assigned high appearance. A compound that also has been observed to bioconcentrate in organisms in nature as a result of OWTS effluent influence is given very high appearance.

Table 2-6. Criteria Characteristics for Top OWCs (of 20 selected contaminants of concern).[1]

Compound	Class	Toxicity/Risk	Persistence in Environment	Appearance in Onsite Wastewater
sulfamethoxazole	antibiotic	moderate	low	low
ibuprofen	pain reliever	very low	high	high
carbamazapine	antiepileptic	low	very high	low
diclofenac	analgesic	high	high	moderate
caffeine	stimulant	very low	high	very high
clofibric acid	lipid regulator	low	moderate	low
gemifibrozil	lipid regulator	low	moderate	low
triclosan	antimicrobial	high	moderate	very high
nonylphenol	surfactant	high	moderate	very high
erythromycin	antibiotic	high	low	moderate to low
bisphenol a	plasticizer	very high	moderate	low
ciprofloxacine	antibiotic	high	very low	low
naproxen	pain reliever	moderate	low	low
1,4-dichlorobenzene	deodorant	high	moderate	high
17-β estradiol	hormone	high	high	moderate
estrone	hormone	high	moderate	moderate
tylosin	antibiotic	high	very low	moderate
testosterone	hormone	high	low	moderate
tetracycline	antibiotic	high	moderate	moderate to low
warfarin	anticoagulant	very high	high	low

[1] Cell is shaded where criteria are met for the characteristic. Compound is shaded where all criteria are met.

The five contaminants chosen for this study (triclosan, 4-nonylphenol, 1,4-dichlorobenzene, 17β-estradiol, and diclofenac) meet the criteria and represent the diversity of compounds considered OWCs. For each of these chemicals, pertinent daughter products will also be considered. A summary of the available literature is provided below for each of these five OWCs. The dominant fate and transport characteristics summarized from this literature review are presented in Appendix A.

2.6.3 Triclosan

Triclosan is an antimicrobial compound found in a variety of household products. It is used in soaps, deodorants, toothpastes, shaving creams, mouth washes, and cleaning supplies. It is increasingly being used in consumer products such as kitchen utensils, toys, bedding, socks, and trash bags. It is found in wastewater, STE, and occasionally river water in the microgram per liter range. Significant concentrations have not been measured in groundwater, although studies that investigate any OWC groundwater contamination are rare. Triclosan is toxic to algal communities in the microgram per liter range (Coogan et al., 2007). Degradation products include methyl-triclosan which is considered environmentally stable under certain conditions, and dioxin, which is highly toxic to organisms including humans and somewhat stable in the environment. Triclosan has also been linked to the formation of antimicrobial-resistant strains of bacteria (Suller and Russel, 2000).

A considerable amount of research has been conducted on the occurrence of triclosan and methyl-triclosan in wastewater, streams, and lakes (Bester, 2005; Böhmer et al., 2004; Coogan et al., 2007; Hua et al., 2005; Lindstroem et al., 2002; Sabaliunas et al., 2003). One study showed the concentration changes of triclosan through treatment units in five wastewater treatment plants. This study observed removal of triclosan by aerobic biodegradation but not by anaerobic biodegradation. Sorption was highly significant with an average of 30% of the aqueous compound sorbing to sludge (Ying and Kookana, 2007). A study by Thompson et al. (2005) revealed the same; removal of triclosan was by aerobic biodegradation and partitioning to the solid phase by sorption to sludge. An investigation of the behavior of triclosan in loamy soil showed that in the experimental conditions, aerobic biodegradation occurred with a half life of 18 days (Ying et al., 2007).

Complete biodegradation does not always occur. Bester et al. (2005) demonstrated the biotransformation of triclosan to methyl-triclosan in the main aeration basin of a wastewater treatment plant. Methyl-triclosan has been shown to be environmentally stable. A study of seven lakes in Sweden revealed the bioaccumulation of methyl-triclosan in fish at a log value of 3.3 to 3.7. The methyl-triclosan concentrations in fish were higher than those of triclosan due to its stability in the environment (Balmer et al., 2004; Lindstroem et al., 2002). Other degradants that may result from aerobic biodegradation and photolysis include 2,4-dichlorophenol, 2,8-dichlorodibenzo-p-dioxin, and oligomerization products (Latch et al., 2005). These degradants were measured in a lab study that may not reflect actual environmental conditions. Volatilization of the compound is predicted to be insignificant based on the low Henry's law constant (air-water partition coefficient). Table 2-7 summarizes triclosan transport parameters reported in the literature. A compilation of fate and transport behaviors can be found in the Triclosan Fact Sheet (Appendix A).

Table 2-7. Chemical Characteristics for Triclosan.

Parameter	Range of Reported Values	Sources
log Koc	4.265 – 4.7	Ying and Kookana, 2007
Solubility (mg/L)	5.0 – 10.0	Seitz et al., 2005
Kh (atm-m3/mole)	1.52E-07	Thompson et al., 2005
log Kow	4.7 – 4.8	DeJong et al., 2004; Ying and Kookana, 2007
Kd	4.3	Samsøe-Petersen et al., 2003
pKa 20°C	8.14	Samsøe-Petersen et al., 2003
Vapor Pressure (Pa) 25°C	7.00E-04	Samsøe-Petersen et al., 2003

2.6.4 4-Nonylphenol

4-nonylphenol is a nonionic surfactant daughter product of nonylphenol polyethoxylates. It is widely used in industrial and commercial detergents, emulsifiers, wetting agents, and dispersing agents. It is found in wastewater, STE, and occasionally river water in the microgram per liter range. No data for groundwater concentrations were found in the literature. 4-nonylphenol is a known endocrine-disrupting compound and is considered toxic to aquatic life (Topp and Starratt, 2000). Because of the high concentrations and ubiquitous appearance of this chemical, the U.S. EPA has established freshwater aquatic life and saltwater aquatic life criteria for nonylphenol. The 1-hour average concentration of nonylphenol cannot exceed 28.0 µg/L more than once every three years and the four-day average cannot exceed 6.6 µg/L more than once every three years (Rand, 1995). This is the only OWC with an aquatic life maximum contaminant level (MCL) set by the EPA.

The fate and transport of 4-nonylphenol has been studied in wastewater treatment plants and soils ranging from sands to clay loams. Hseu (2006) demonstrated the dependence of biodegradation rate on initial mass of 4-nonylphenol and soil type. The study found that high rates of biodegradation occur in sandy loams, which suggests that anaerobic degradation is more effective than aerobic because loam soils generally have lower oxygen diffusion rates and higher organic carbon content. Higher rates of biodegradation were associated with higher initial concentrations of 4-nonylphenol. Sorption of 4-nonylphenol to the organic carbon fraction of the soil is found to be high (Ahel and Giger, 1993; Ahel et al., 1994). Low biodegradation rates for 4-nonylphenol are shown to increase with increased soil respiration (Roberts et al., 2006). This same study showed that the plants in the rhizosphere have no significant impact on nonylphenol mineralization though removal is temporarily achieved by plant uptake and rates of removal are highly dependent on pH and temperature.

Information on aqueous concentrations of 4-nonylphenol is not complete without analyzing for the degradants 4-tert-octylphenol, 4-tert-octylphenolmonoethoxylate, 4-tert-octylphenoldiethoxylate, 4-tert-octylphenoltriethoxyalte, and 4-tert-octylphenoltriethoxylate (Das et al., 2004; Conn et al., 2006). Abiotic degradation in terms of photolysis has been shown to be important in lake water. In the summer at noon with clear skies, the half life measured due to photolysis is 10 to 15 hours (Ahel et al., 1994). Volatilization of the compound is predicted to be insignificant based on the low Henry's law constant. Table 2-8 summarizes 4-nonylphenol transport parameters reported in literature. A compilation of fate and transport behaviors can be found in the 4-nonylphenol Fact Sheet (Appendix A).

Table 2-8. Chemical Characteristics for 4-Nonylphenol.

Parameter	Range of Reported Values	Sources
log Koc	4 - 5.9	Ahel and Giger, 1993; McLeese et al., 1981
Solubility (mg/L)	4.6 - 11.9	Ahel and Giger, 1993
Kh (atm-m3/mole)	2.45E-09	Hellman 1987
log Kow	4.2 - 4.48	Ahel and Giger, 1993; McLeese et al., 1981
Kd	80-120	Duering 2002
pKa 20 °C	10.7	McLeese et al., 1981
Vapor Pressure (atm) 25 °C	1.00E-06	Ahel and Giger, 1993

2.6.5 1,4-Dichlorobenzene

1,4-Dichlorobenzene is a disinfectant, deodorant, and insect repellant. It is found in many household cleaners, particularly toilet-bowl cleaners. It is a possible carcinogen, neurotoxin, and endocrine disruptor. Concentrations in raw wastewater and STE are lower than those of triclosan and 4-nonylphenol but are still in the microgram per liter range. No data for groundwater or stream water contamination related to wastewater sources were found in the literature.

The fate and transport of 1,4-dichlorobenzene has been studied in onsite wastewater treatment, wastewater treatment plants, groundwater, and in two laboratory soil studies. One soil study investigates behavior in a sandy soil with a 0.087 fraction of organic carbon. A retardation factor of 3.4 to 3.7 was calculated in a soil column by Wilson et al. (1981). 1,4-dichlorobenzene is more mobile in soils that are organic carbon-poor (Schwarzenbach and Westall, 1981). Biological degradation was not considered a significant removal mechanism in the Schwarzenbach study. Conversely, in another study, aerobic biodegradation resulted in removal of the compound with a half life of 15 days (Robertson, 1994). Anaerobic degradation was not observed. Soil type or class information for this study was not reported. Aqueous-phase mass loss due to volatilization of the compound was a dominant process in both of these soil studies. A study that investigated the long term fate of 1,4-dichlorobenzene in groundwater demonstrated that when the system is depleted of dissolved organic carbon (DOC), no more degradation will occur (Barber et al., 1988). This study also suggests that retardation due to sorption is not significant. Table 2-9 summarizes 1,4-dichlorobenzene transport parameters reported in the literature. A compilation of fate and transport behaviors can be found in the 1,4-Dichlorobenzene Fact Sheet (Appendix A).

Table 2-9. Chemical Characteristics for 1,4-Dichlorobenzene.

Parameter	Value Range	Sources
log Koc	2.44	Chiou et al., 1983
Solubility (mg/L)	80	Yalkowsky 2003
Kh (atm-m3/mole)	2.41E-03	Bahadur et al., 2007
log Kow	3.44	Hansch 1995
Kd	--	--
pKa 20°C	--	--
Vapor Pressure (mm hg) 25°C	1.76	ATSDR 1998

2.6.6 17β-Estradiol

17β-estradiol is an endocrine disrupting compound and the most persistent of the hormone compounds. It is used as a hormonal contraceptive and for hormone-replacement therapy in blocking hormones to treat hypoestrogenism. 17β-estradiol degrades primarily to estrone. Concentrations are much lower than the other OWC compounds in wastewater, STE, and rivers (in the hundredths of nanogram per liter range). These low concentrations are still enough to contribute to "gender-bender" effects in fish and amphibians as described earlier in this report.

A few studies have been conducted on the behavior of 17β-estradiol in soils receiving STE from OWTSs. 17β-estradiol is shown to be the most persistent of the common hormones (Das et al., 2004; Mansell and Drewes, 2004). In OWTS, sorption to sludge in the septic tank has been reported to be the most important removal mechanism followed by aerobic biodegradation (Drewes et al., 2005; Yu et al., 2004). According to Mansell et al. (1994), adsorption to soil was a primary removal mechanism as it allowed for increased time for biodegradation, but the organic carbon composition of the wastewater had no significant effect on the amount sorbed. This study also found that of the initial concentration of 17β-estradiol in treated effluent that was discharged to the soil, 25% was detected in a shallow monitoring well located at a depth of 5.1 m with an approximate travel time of one day. Published half lives for 17β-estradiol range from 0.8 to 9.7 days depending on soil type (Lee et al., 2003). The shorter half life corresponds to a sandy soil while the longer half life was from a clay-rich soil study. Volatilization of the compound is predicted to be insignificant based on the low Henry's law constant. Table 2-10 summarizes 17β-estradiol transport parameters reported in literature. A compilation of fate and transport behaviors can be found in the 17β-Estradiol Fact Sheet (Appendix A).

Table 2-10. Chemical Characteristics for 17β-estradiol.

Parameter	Value Range	Sources
log Koc	3.14 - 4.09	Das et al., 2004; Yu et al., 2004
Solubility (mg/L)	3.1 - 13	Yu et al., 2004; Lee et al., 2003; Mansell et al., 2004
Kh (atm-m3/mole)	2.64E-11	Mansell et al., 2004
log Kow	3.94 – 4.1	Yu et al., 2004; Lee et al., 2003; Mansell et al., 2004
Kd	3.56 – 83.2	Das et al., 2004
pKa 20 °C	10.23	Yu et al., 2004
Vapor Pressure (mm hg)	--	--

2.6.7 Diclofenac

Diclofenac is a non-steroidal anti-inflammatory drug (NSAID) taken to reduce inflammation and as an analgesic reducing pain in conditions such as arthritis or acute injury. It can also be taken to reduce menstrual pain. At very low concentrations this chemical has been shown to cause renal failure resulting in immediate death in birds such as vultures (Oaks et al., 2002). Concentrations in wastewater and STE are in the microgram per liter range and in the nanogram per liter range for river water. No data for groundwater contamination related to wastewater sources were found in the literature. Diclofenac bioaccumulates and has the potential to reach higher concentrations in river fish, which are a food source for some birds of prey. Thus, avian species may be at risk.

Most fate and transport studies for diclofenac focus on photolysis because it is a dominant removal mechanism. A study of the behavior of diclofenac in 12 wastewater treatment plants demonstrated the persistence of the compound through these systems. Overall elimination rates for the treatment plants and lab studies that simulated treatment plant processes were very low (Strenn et al., 2004). This suggests that anaerobic biodegradation rates, sorption to organic carbon, and abiotic degradation rates are very low for diclofenac. Volatilization of the compound is predicted to be insignificant based on the low Henry's law constant.

Diclofenac is quickly and effectively removed by indirect and direct photolysis. However, removal rates in wastewater treatment remain low because onsite treatment processes rarely include direct access to sunlight. Use of UV lamps in onsite systems may increase treatment. A study by Perez-Estrada et al. (2005), investigated various photolytic degradation pathways. This study showed that in photolysis, diclofenac transformation is mainly due to the initial photocyclization of diclofenac into carbazole derivatives. Ultimately the study showed that diclofenac follows different reaction pathways depending on the degradation treatment. Another study showed that elimination by direct phototransformation has the highest first-order rate (0.088 day^{-1}) followed by sedimentation (0.005 day^{-1}) and indirect phototransformation (0.001 day^{-1}) (Tixier et al., 2003). A study by Buser et al. (1998) measured the rates of photodegradation in the summer versus the winter (0.081 day^{-1} and 0.022 day^{-1} respectively) and analyzed the lake sediment to ensure that all diclofenac loss was from photolytic decay. Table 2-11 summarizes diclofenac transport parameters reported in literature. A compilation of fate and transport behaviors can be found in the Diclofenac Fact Sheet (Appendix A).

Table 2-11. Chemical Characteristics for Diclofenac.

Parameter	Value Range	Sources
log Koc	2.43 - 3.87	Scheytt et al., 2006
Solubility (mg/L)	2.4	Scheytt et al., 2004
Kh (atm-m3/mole)	--	--
log Kow	4.51	Scheytt et al., 2004
Kd	0.8 - 5.9	Beausse 2004
pKa 20 °C	3.99 - 4.16	Scheytt et al., 2004; Escher et al., 2006
Vapor Pressure (mm hg) 25 °C	6.14E-08	Scheytt et al., 2006

2.6.8 Discussion

Future research is imperative for a better understanding of fate and transport processes of organic wastewater contaminants as they are integrated into the environment, particularly with respect to removal in STUs associated with OWTS. The most relevant removal processes for onsite systems are sorption, biodegradation, and volatilization. The soil-water partitioning coefficient (K_d) needs to be measured for a variety of soil types and quantitative relationships should be developed between K_d and the fraction of organic carbon in the soil. It is known that organic molecules partition to organic carbon in the soil, however, there is yet to be a study clearly investigating and quantifying this relationship for the selected OWCs. In addition, it is important to know whether sorption isotherms are linear or nonlinear, whether sorption is an equilibrium or kinetic process, and whether or not sorption is reversible. Reversible sorption does not remove mass from the soil water, but causes the transport of the contaminant in the soil to be retarded (slowed). This can allow more time for degradation processes to occur. Irreversible sorption can remove chemical mass from the aqueous phase (e.g., treatment), but the mechanisms causing such sorption, and the soil-water geochemical conditions that are necessary, must be understood. If sorption isotherms are nonlinear, then there would be an effective maximum sorption capacity of the soil, so this attribute is critical to study. Rates of volatilization for chemicals with relatively high Henry's law constants must be assessed for unsaturated soil conditions. Microbial degradation can remove contaminants, but we must understand whether degradation occurs under aerobic or anaerobic conditions (or both), and also understand the degradation pathways. Any toxic degradation daughter products, and their transport characteristics, must also be examined. Little research has been conducted in these areas related to volatilization, sorption, and biodegradation.

Because sorption to soil or sludge and biodegradation are so critical in the removal of these compounds from the aqueous phase, it is likely that the biozone on and just below the infiltrative surface of an STU plays an extremely important role in the fate and transport of OWCs. The biozone is a region of increased organic carbon and microbial development at the infiltrative surface and a zone of increased microbial activity and byproducts in the soil below the infiltrative surface. This zone could result in increased sorption and biodegradation, and the role of the biozone in transformation these compounds needs study.

With an appropriate conceptual model for the transport and removal characteristics in soil including descriptions of behaviors using partitioning coefficients, sorption rate coefficients and degradation rate constants, the expected contaminant removal with depth for certain loading rates, soil types and conditions can be predicted and/or measured for a certain soil. This can be accomplished by conducting field and laboratory experiments combined with numerical modeling. A brief discussion is provided below of STU features or factors that may result in optimum removal based on compound characteristics.

2.6.8.1 Predictions for the Most Effective Soil Treatment Unit Design

Based on previous studies of fate and transport characteristics of the five priority organic compounds, some preliminary conclusions can be drawn concerning the most effective STU design to remove these contaminants from the STE infiltrating through soil (Table 2-12). Soil treatment unit types include infiltration trenches (standard and shallow), infiltration beds, and seepage pits. Infiltrative systems can be installed "at-grade" (i.e., just below the natural ground surface), or within raised mounds, which are used when soil conditions are not conducive to infiltration or the seasonal water table is very close to the ground surface. Standard trenches have a typical infiltrative surface depth of 60 cm (2 ft) below ground surface (bgs); shallow trenches have a typical infiltrative surface of about 15 cm (6 in.) bgs. Beds, which have wide rectangular or square geometry, are not commonly used because of oxygen exchange issues leading to higher failure rates and lower performance. Seepage pits are deep circular excavations that rely almost entirely on sidewall infiltration. They are no longer permitted in many places because of their depth and small horizontal profile. This creates a greater point-source pollutant loading potential to groundwater.

Table 2-12. Summary of Predictions of Most Effective STU Design and Dosing Method for Each Contaminant.

Contaminant	STU Geometry	Effluent Delivery Method
Triclosan	shallow trench	periodic dosing
4-Nonylphenol	standard trench	periodic dosing
1,4-Dichlorobenzene	shallow trench or mound	pressurized spray dispersal
17β-Estradiol	trench (shallow or standard)	continuous dosing
Diclofenac	shallow trench/FWS wetland before STU	continuous dosing

Dosing method is another feature to be considered in STU design. With timed dosing, pumps deliver effluent on a timed cycle. This results in relatively dry and wet periods which influence soil moisture, aeration, and redox condition in the soil over the course of a day. Continuous dosing involves constant delivery of effluent at a low rate. Constant addition of effluent may result in an enhanced biomat and consequently lower infiltration rates compared to the saturated hydraulic conductivity. This increases volatilization and gas-phase diffusion for oxygen and possibly the more volatile OWCs. The increase in oxygen may promote aerobic and some forms of abiotic degradation while inhibiting anaerobic degradation. On-demand dosing delivers effluent when the dose tank reaches a certain volume. The pumps will turn off when the volume is lowered to a certain point. On-demand dosing results in changing soil-water conditions as well, but the time intervals between when the pump is on and when it is off are likely to be longer than those for timed dosing. Mode of effluent delivery is another important distinction between STU designs. A drip dispersal method delivers effluent through the trench or across the mound equally over time. A chamber can be used in a trench rather than using gravel fill to stabilize the trench and increase contact with air. Pressurized spray distribution results in more uniform distribution of applied effluent to the infiltrative surface as well as increased contact with air in the trench chamber.

Triclosan's dominant removal mechanisms (for aqueous concentrations) are sorption to the carbon fraction of the soil and aerobic biodegradation (Section 2.6.3). Shallow trench dispersal will allow for interaction with the rhizosphere where plant uptake and increased bacteria activity may assist removal. Periodic dosing or lower hydraulic loading rates applied over a larger area will promote aerobic conditions. A trench system may also allow maximized biozone development because of the increased infiltrative surface area. Based on results in a sandy loam, removal is expected in the top 1.5 of the soil profile (Conn, 2008).

Nonylphenol's dominant removal mechanisms, similar to triclosan, are sorption and aerobic biodegradation. Trenches maximize the infiltrative surface area where the biozone is facilitated and effectiveness of treatment may be increased. Effective effluent distribution methods should be similar to those for triclosan because they best accommodate aerobic biodegradation.

The dominant aqueous phase mass removal mechanisms for 1,4-dichlorobenzene include sorption to soil and volatilization (Section 2.6.5). Volatilization is an important process for partitioning the compound out of the aqueous phase where it can be more rapidly transported to the atmosphere (gas diffusion rates in soil can be very high), and partitioning to the gas phase can also retard aqueous phase transport resulting in more time for biodegradation to occur. Thus, maximized contact with air is essential for removal of 1,4-dichlorobenzene. Pressurized spray dispersal into a trench system will provide most air contact and thus the best removal rates. Evapotranspiration may also aid in removal. In this case, a shallow trench or a mound system would likely provide the most potential for evapotranspiration. Application methods that result in overall decreases in the average water content of the unsaturated soil also will enhance volatilization. Thus, dispersing at lower rates over a wider area, or intermittent dosing, may enhance removal. However, anaerobic degradation would be inhibited for any operating condition that promotes increased air content and volatilization.

Sorption to soil is the dominant mode of removal of aqueous mass of 17β-estradiol (Section 2.6.6). Because the trench system maximizes surface area of soil exposure, it is likely the best STU geometry for this contaminant. Sorption for 17β-estradiol is likely increased in the organic carbon rich biomat. If the continuous dosing technique does contribute to a more substantial biomat, it is likely that continuous dosing will maximize treatment. However, understanding whether sorption is an equilibrium or kinetic process, reversible or irreversible, linear or nonlinear, is critically important. If sorption is reversible, it may not be a removal mechanism; however, it may slow transport increasing time for biodegradation to occur.

Diclofenac is mainly degraded by photolysis in the presence of sunlight, a condition that is unlikely in an STU. A shallow trench, however, provides access to the rhizosphere where plant uptake may be a removal process. Studies to investigate this hypothesis should be completed before any assumptions can be made. A free-water surface constructed wetland is an onsite wastewater treatment unit that allows for contact with the atmosphere. Photolytic degradation is most likely to occur if the treated wastewater is routed through one of these units before being integrated into the environment via an STU. Because abiotic degradation may have a significant contribution to the removal of diclofenac as the molecule interacts with oxygen, higher oxygen gas phase diffusion may result in increased treatment. Continuous dosing may accommodate this as water content directly below the biomat remains low and air content remains high.

2.6.8.2 Importance of Infiltration Surface Biomat

Many OWCs (including triclosan, nonylphenol, and 17β-estradiol) have been shown to sorb to and biodegrade in sludge (Bester, 2005; Carballa et al., 2004; Castiglioni et al., 2006; Jones et al., 2006). Sludge is rich in organic carbon and hosts a large and diverse population of bacteria. The biomat that forms on the infiltrative surface in an STU has these same features; it is likely that sorption and biodegradation are important for OWC transformation and removal in the soil profile. This zone only extends through the first several centimeters of soil but reduced hydraulic conductivity, increased fraction of organic carbon, and increased bacterial activity suggest this may be a controlling layer for the fate and transport of OWCs. This hypothesis requires further experimental investigation to better characterize the effect of the biomat zone.

2.6.8.3 Next Steps

Clearly, insufficient data exists in the literature to test the preliminary conclusions drawn above. Thus, to test these hypotheses, the numerical model, HYDRUS 2-D and 3-D, may be coupled with known transport parameters to estimate removal with depth for certain loading rates and soil types under certain site conditions. Field and laboratory experiments investigating contaminant transport of a compound with time and depth under specific conditions will allow for more accurate assessment of certain processes as well as estimation of removal rates, and thus enable parameterization of a numerical model. It is not cost-effective to conduct field or laboratory experiments for all five contaminants for all relevant OWTS/STU operating conditions. Thus, limited laboratory experiments should be conducted, and mathematical modeling should be used to analyze the data from these experiments. Once the models are shown to accurately simulate contaminant transport and transformations for the selected operating conditions, then the models can be used to test a much broader range of operating conditions, and the impact on OWC contaminant removal. Results from Conn's (2008) study will help to direct which laboratory tests may be most critical.

At CSM, samples can be analyzed for aqueous concentrations of these OWCs using solid phase extraction techniques with gas chromatography/mass spectrometry (GC/MS). Compounds, as well as added surrogates and standard mixes of known concentration, are analyzed using target compound identification based on SIM (selected ion monitoring) analysis with calculated retention times (+/- 0.05 min) for compounds identified using ion ratios (3 ions +/- 20%). An external calibration curve that is normalized to the internal standards is used to quantify concentrations of compounds in the sample.

CHAPTER 3.0

Review of Available Modeling Tools

3.1 Introduction

Since the advent of powerful personal computers, great progress has been made in simulating the transport and fate of contaminants in the subsurface. Basically, there are two types of models: analytical and numerical. Analytical models provide exact solutions whereas numerical models yield approximate solutions. Models to simulate, one-, two- and three-dimensional domains have been developed. Because many processes must be approximated in analytical solutions, numerical models are typically applicable over a wider range of environmental conditions, i.e., for the simulation of processes in heterogeneous flow and transport domains. The decision on what model to use must be based on available data, the complexity of the site and the accuracy of the information required. In additions, assumptions within each model must be known to correctly apply the simulation results (Mulligan and Yong, 2004).

Many models have been developed for N agricultural applications but very few have been developed for OWTS and the processes that occur in the STU. Similarly, a number of models have been developed for P fate and transport but again very few have been developed for OWTS and the processes that occur in the STU. Modeling microbial matter has mostly focused on transport of viruses, which are the smallest particles relative to bacteria and protozoa, with modeling approaches to microbial transport greatly varied depending on the scale of the study and the specific interests of the investigators. Finally, several studies have been conducted to model fate and transport of OWCs in the environment. In this chapter a review of the existing OWTS models is presented with discussion that could guide development of an OWTS model.

3.2 Modeling Nitrogen in Onsite Wastewater Treatment Systems

Addiscott and Wagenett (1985) suggested a framework for classifying models into deterministic and stochastic approaches. Most N models are deterministic in that they do not incorporate the effect of spatial variability in their predictions. However, a few stochastic models have been developed to account for the large variability in micro-site denitrification rates (Parkin and Robinson, 1989; Arah, 1990). They further classify models as mechanistic or functional. Functional models usually use mass balance or capacity functions to fill and deplete pools or soil layers, whereas mechanistic models deal in rates or differential equations. Mechanistic models can be divided into numerical or analytical approaches, depending on how the rate equations are solved.

Another way to classify models is in terms of spatial and temporal scales (Shaffer, 1995). Models can be pedon, field, farm, watershed, or regional scale. Most N models are field scale. Models can use time steps that range from seconds or less for numerical models to years for mass balance models such as the CENTURY model (Metherell et al., 1993).

A number of reviews of N models have been published. These include Shaffer (1995), Valiela et al. (2002), Beal et al. (2005), and McCray et al. (2005). Of the reviews that included OWTSs, Valiela et al. (2002) reviewed models that simulated groundwater loading of N to estuaries where the contribution from OWTS was considered important. Beal et al. (2005) reviewed OWTS with an emphasis on Australian systems and suggested that models of single-pass sand filters should be explored to understand treatment in OWTS biomats and soils. McCray et al. (2005) reviewed the literature for model parameters suitable for simulating the fate and transport of N and P in OWTS. They found that there were relatively few studies that quantified denitrifcation rates specifically for OWTS and concluded that the biggest question in modeling N in OWTS is under what conditions and to what extent does denitrification occur.

3.2.1 Nitrogen Transformations

The most common forms of N in STE are organic N and NH_4^+ (McCray et al., 2005). In modeling fate and transport of N within the STU, the N transformations of interest (Figure 2-1) are:

- the transformation of organic N to NH_4^+, which is called *ammonification*, *hydrolysis*, *mineralization* (which can also include the transformation to NO_3^-), or *decomposition*
- the transformation of NH_4^+ to NO_3^-, which is called *nitrification*
- the transformation of NO_3^- to N_2O gas or N_2 gas, which is called *denitrification* (Brady and Weil, 2008)

Ammonification can occur under aerobic or anaerobic conditions so this transformation proceeds within the septic tank and most of the N in STE is in the form of NH_4^+ (75% with the rest as organic N) (McCray et al., 2005). The bacteria responsible for nitrification (a two-step process) are more sensitive to environmental conditions than those involved in ammonification. They require aerobic conditions, moisture, and a source of carbon (C). Plant uptake of NH_4^+ and NO_3^- can occur. These forms of N can also be taken up by microbes and converted back to organic N through *immobilization*, the reverse process of mineralization. Ammonium as a cation is adsorbed by the cation exchange capacity (CEC) of soil but NO_3^- is highly mobile. Ammonium is subject to volatilization losses as ammonia gas.

The bacteria responsible for dentrification are the most sensitive to environmental factors. Tiedje (1988) proposed a conceptual model for the environmental regulation of denitrification in soils (Figure 3-1). Environmental factors rather than the lack of organisms are the most likely reasons for a reduction in denitrification. Extreme temporal and spatial variability has been found in denitrification rates, most likely due to local variation in environmental factors. Denitrifiers are aerobic bacteria which have the alternative capability to reduce N oxides when oxygen is not available. Oxygen availability is determined by the difference between the rate of supply (affected by diffusion) and the rate of consumption (respiration). The main environmental factors directly affecting denitrification are O_2, NO_3^-, and C fluxes to the site of denitrification (Figure 3-1). Several environmental factors (water, plants, and OM) have multiple indirect effects. Carbon (OM) has three potential effects: a direct effect as a source of energy (electron donor) (bottom Figure 3-1), an indirect effect as a source of NO_3^- (middle Figure 3-1), and an indirect effect in driving respiration which consumes oxygen (top Figure 3-1). Of the three effects, the role of carbon in driving the demand for oxygen and producing micro-sites of low oxygen availability or hotspots of denitrification is probably the most important. Soil water content affects the diffusion of oxygen, NO_3^-, and C and the rate of respiration. By freezing and thawing, water also disrupts aggregates and increases the availability of C. Soil texture, plant uptake, and rainfall affect soil water content and contribute to the spatial and temporal variability of this environmental factor. Diffusion coefficients for oxygen are four orders of magnitude less in soil water than in air (10^{-7} vs. 10^{-3} cm^2/s). The supply of NO_3^- is usually not a limiting factor in agricultural topsoils, but can be important in forest soils. In constantly anaerobic habitats, the supply of NO_3^- can become limiting due to limited nitrification.

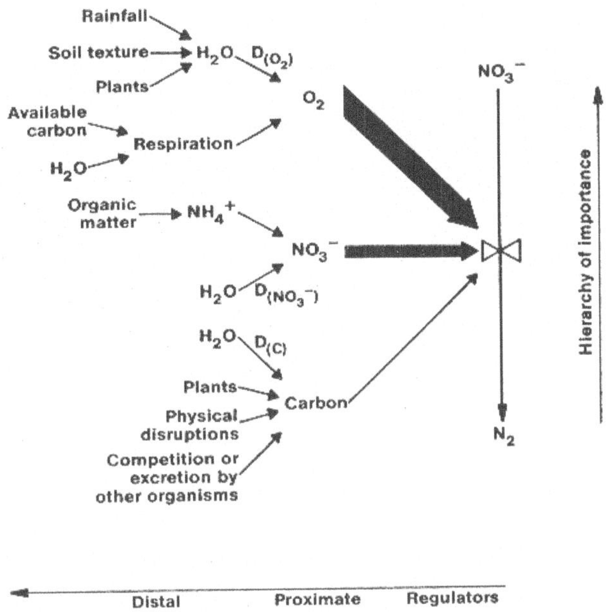

Figure 3-1. Conceptual Model of Environmental Regulation of Denitrification.
The vertical dimension shows the hierarchy of importance of the three major regulators of denitrification. The horizontal dimension illustrates the proximity of the regulatory factors or features to the enzymatic process. D refers to an effect on diffusion rates of the indicated regulator. Tiedje, 1988. Reprinted with permission from John Wiley & Sons, Inc.

3.2.2 Plant Uptake

Potentially as important as N transformations, plant uptake may play a role in the fate and transport of N in OWTSs, but there is limited information on modeling this component. Simple onsite wastewater treatment N models, such as those described in the next chapter, generally assume that plant uptake of N is not important in N assimilation. More complex models, such as those described in later chapters, do account for plant uptake of N. Subsurface drip irrigation systems (SDIS) may maximize plant uptake of N and be especially suitable for marginal sites that have low permeability soils, difficult installation areas, or high water tables (Bohrer, 2000). The concern is what might happen with these systems during winter when plants are dormant or transpiring at minimal rates (Ruskin, 1992). It will also be important whether or not plant residues such as grass clippings are removed from the site. The EPA manual for land application systems (LASs) provides guidance for estimating plant uptake of N for slow rate LASs and this may be applicable to SDIS (U.S. EPA, 2008b). Slow rate systems are designed to minimize runoff and rely on plant uptake of N as one of the main components of N assimilation. They also rely on a consistent cycle of wet and dry conditions to maximize infiltration and N transformations. The design of a slow rate LAS depends on the limited design parameter (LDP). For slow rate systems, the LDP is usually N or infiltration capacity, which is similar to the design of OWTS. Optimum design of a slow rate LAS usually results in the selection of perennial grasses because of a longer application season, higher infiltration capacity, and greater N assimilation compared to other annual crops. In the EPA (2008b) manual for LAS, N removal in vegetation can be estimated based on dry matter production, harvest index, and percent N of the harvested vegetation.

3.2.3 Onsite Wastewater Treatment System Nitrogen Models

Very few models have been developed for fate and transport of N in OWTS. Most that exist are simple models designed to predict the NO_3^- concentration in groundwater beneath subdivisions using OWTS due to dilution effects. However, three models (Beggs et al., 2004; Hassan et al., 2008; Heatwole and McCray, 2007) have been developed using the HYDRUS model framework (Šimůnek et al., 2005, 2006) and these are discussed in Section 3.2.5 on HYDRUS models.

One such model is that developed by Hantzsche and Finnemore (1992) for subdivisions in California. Using a mass balance approach, they assumed that the NO_3^--N concentration in recharge water reaching the groundwater (n_r) [M L^{-3}] was a weighted average of the concentration in wastewater (n_w), adjusted for expected losses due to denitrification, and the concentration in background rainfall recharge (n_b), where the weights were the wastewater recharge rate (I) and the rainfall recharge rate (R) [L T^{-1}]:

$$n_r = \frac{I(1-d)n_w + Rn_b}{I+R} \qquad (3.2.3\text{-}1)$$

The fraction of NO_3^--N lost to denitrification is d [-]. Dividing by R yielded a relationship between n_r and I / R shown in Figure 3-2. Two model curves are shown, one for no fractional loss to denitrification ($d = 0$) and one for a commonly assumed fraction ($d = 0.25$). The groundwater median concentrations of NO_3^--N measured in three subdivisions (the Bolinas Mesa site included a north and south area of the subdivision) and associated values for I/R are also shown. If the assumptions in the model were correct, the figure shows that denitrification losses were somewhere between 0 and 25%. The assumptions in the model included: 1) an effluent NO_3^--N concentration (n_w) of 40 mg/L, 2) no dispersion, 3) no fertilizer input, 4) complete mixing between wastewater recharge and groundwater, and 5) no up-gradient groundwater flow. Figure 3-2 shows the important effect that recharge rate (R) has on the concentration of NO_3^- in groundwater. The authors developed an equation based on this model to predict the area that would be required for each house to maintain groundwater NO_3^- concentrations below a concentration set by several regional authorities in California to determine appropriate housing densities.

A similar approach was used by Taylor (2003) to develop an equation that predicted the area required for each house in subdivisions in Pennsylvania. Their model included the input from fertilizers and up-gradient groundwater lateral flow. They assumed no losses due to denitrification (site specific information was used to change the default value). The model was implemented as a computer spreadsheet program.

The New Jersey Office of State Planning also developed a simple groundwater loading model to predict acceptable housing densities (NJOSP, 1988). Aquifers in sandy coastal plain soils in the New Jersey Pinelands have a limit of 2 mg/L NO_3^- to protect the ecosystem in this region. The model did not include up-gradient groundwater flow because the authors assumed that there was little mixing of NO_3^- plumes with ambient groundwater. No dentrification within the unsaturated zone or groundwater was assumed. The average NO_3^- concentration for input from OWTS effluent used in the model was 40 mg/L.

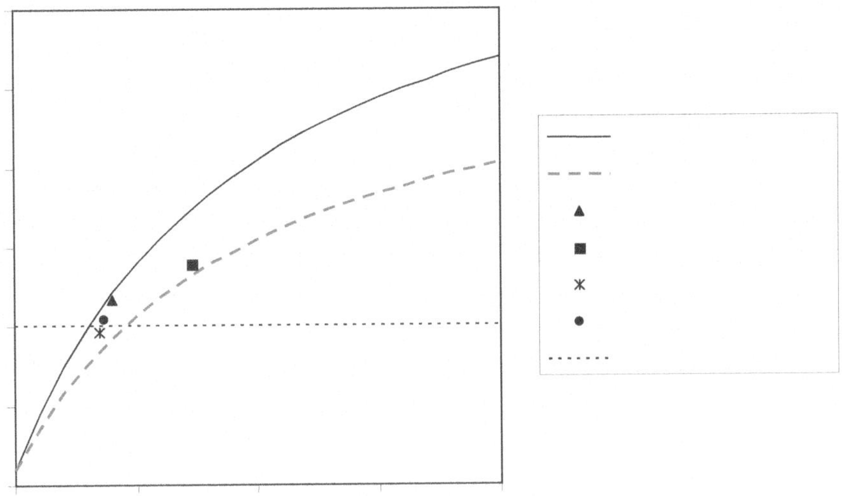

Figure 3-2. Comparison of Measured and Predicted Groundwater NO_3^--N Concentration in Four Subdivisions in California. Adapted from Hantzsche and Finnemore, 1992. Reprinted with permission.

Frimpter et al. (1990) developed a similar groundwater loading model for use in Massachusetts to protect public drinking wells from NO_3^- contamination. The capture zone of a well was defined based on aquifer hydraulic properties. Within the capture zone, an equation similar to 3.2.3-1 was used to estimate the NO_3^- concentration in the groundwater reaching the well based on land use within the capture zone, including homes on OWTS. The model assumed that there were no losses to denitrification in the unsaturated zone or in groundwater. Losses were due entirely to dilution. The model results showed that increasing the pumping rate of the well expanded the capture zone and could change the concentration of NO_3^- reaching the well, depending on land use in the expanded capture zone.

Eichner et al. (1992) developed a mass-balance groundwater loading model for use in Cap Cod, MA. Their model assumed no denitrification in the unsaturated zone or groundwater. The average concentration for loading to groundwater from OWTS used in the model was 35 mg NO_3–N/L. Nitrogen lost from fertilization of lawns was included in the model.

Weintraub et al. (2004) developed a modification to the Watershed Analysis Risk Management Framework (WARMF) model that simulated the fate and transport of N in soil beneath an OWTS. WARMF is a watershed-scale, mass balance model, used to predict nutrient loading to surface water bodies. The module simulated the development of a biozone or biomat (2 cm in thickness) and the changes in hydraulic properties that occurred as the biozone matured. The model included biochemical oxygen demand (BOD) decay, nitrification and adsorption of NH_4^+, and denitrification. The module was calibrated using sand column data (Siegrist et al., 2002), adjusting the parameters for biozone hydraulic properties and decay rates for N species. The module was then tested in predicting the time to hydraulic failure of field-scale OWTS. The full model was applied to the Dillon Reservoir Watershed in Colorado where approximately 1500 OWTSs were present. The model predicted that a scenario for converting from OWTS to a centralized sewer system would increase the total N load to surface waters due to the increased contribution of the waste water treatment facility when the soil's treatment capacity was bypassed.

Several models have been developed for N fate and transport in groundwater where the source of N is OWTS. Anderson et al. (1998) developed a model for a hypothetical subdivision of 20.2 ha (50 acres), housing densities of 5 and 10 homes per ha, and NO_3^--N concentration in effluent entering groundwater of 50 mg/L. Groundwater solute transport was simulated using the convection dispersion equation (CDE), assuming steady groundwater flow in one direction with longitudinal dispersion. They chose not to include denitrification in order to make a conservative (worst-case scenario) estimate of the migration of NO_3^-. An analytical solution to the CDE was used to plot contours of NO_3^- concentrations down-gradient of the model subdivision. They used the model to predict the extent of plumes after five years in two shallow aquifer materials characteristic of Florida: Region 1 with high groundwater dispersivity (18.3 m), and a low mean velocity (0.037 m/day) and Region 2 with a low groundwater dispersivity (9.2 m) and a high mean velocity (0.366 m/day). In Region 1, NO_3^--N concentrations in groundwater were above 10 mg/L everywhere within the subdivision for both housing densities. With five homes/ha, the 10 mg/L NO_3^--N contour had not moved beyond the subdivision area; with 10 homes/ha, the plume extended about 30 m beyond the area. The limited movement of the plume outside the subdivision was attributed to the relatively low groundwater velocity, despite the high value for dispersivity. In Region 2, with five homes/ha, concentrations in the groundwater did not reach 10 mg/L NO_3^--N anywhere, even within the subdivision area. The leading edge of plume moved far beyond the subdivision area, but the concentrations were low. With 10 homes/ha, concentrations above 10 mg/L appeared beneath half the subdivision and the plume extended well beyond the area. Although dispersivities were low, the higher groundwater velocity caused more dispersion of the plume. The study showed that aquifer transport characteristics are important in predicting groundwater concentrations of NO_3^--N.

Another groundwater N fate and transport model for OWTS is that developed by Spiteri et al. (2007). They modeled groundwater plumes from two OWTS sites in Canada. One site had a groundwater system high in Ca (Cambridge site) and the other high in Fe oxides (Moskoka site). They used a one-dimensional numerical CDE model to simulate horizontal transport in the groundwater along the center of the plume using about 15 years of transect well data at both sites. They simulated 15 solutes including N, P, and DOC, and 21 reaction pathways. They assumed there were two fractions of DOC: a labile fraction in the sewage effluent and a less labile background fraction in the groundwater. Ammonium sorption was described by a linear isotherm. Some of the reaction parameters came from the literature and others were fitted using the model predictions of well concentration data. Both groundwater systems were considered oxic (oxygen concentrations above 3 μM). Although up to 90% of the DOC was removed through degradation in the STU unsaturated zone, the remaining DOC was still a significant input to groundwater. The DOC and dissolved oxygen (DO) concentrations decreased with distance in the well data indicating aerobic degradation of DOC within both plumes and oxidation of Fe^{2+}at the Muskoka site. At the Cambridge site, the pH of the plume was near neutral due to the presence of $CaCO_3$, but at the Moskoka site the pH dropped to 4.9 in the plume. Although N concentrations decreased with distance in the plumes, model results showed that decline was due entirely to dilution and that denitrification was limited by DOC availability.

Kinzelbach et al. (1991) developed a groundwater model of denitrification that was designed for natural groundwater systems and groundwater remediation projects. In natural groundwater systems, mineralizable OM (DOC) is usually the limiting factor. In groundwater remediation projects where the objective is to degrade an organic pollutant, NO_3^- can be limiting.

Their model was a multi-chemical species model that simulated transport of oxygen, NO_3^-, organic C, and microbial mass in two dimensions. They showed two comparisons to field data: one based on a natural system and another based on a remediation project. They modeled three phases: a solution phase, an aquifer matrix (solid) phase, and a "biophase". The biophase could be sorbed bacteria. As a separate phase, the model allowed a diffusion-limited exchange between the biophase and the solution phase. This could correspond to aggregate microsites similar to the model of Arah (1990) (described below in Section 3.2.4) or diffusion between different large-scale zones within an aquifer. Oxygen and NO_3^- entered the aquifer through recharge and from the gas phase at the water table. Organic C could come in via recharge but it could also be present in the aquifer matrix (this would correspond to C hotspots in soil).

Kinzelbach et al. (1991) modeled bacterial growth with a Michaelis-Menten equation. To guarantee the possibility of microbial growth in each cell of the model, a small number of bacteria were always conserved. A time-series plot of model predictions for a 1-D column showed that DO dropped first, followed by a decrease in DOC and later NO_3^-. Once DOC reached zero, there was no further reduction in NO_3^- (the system became C-limited). Model predictions of a natural aquifer system were compared to data from a site in Germany sampled over two years. Three forms of the model were used: 1) the only source for C was recharge, 2) the only source for C was recharge and diffusion-limited exchange between the solution and biophase was implemented, and 3) C came from recharge and the aquifer matrix. With the first model, bacterial growth and denitrification occurred only at the top of the aquifer (they depleted the C and/or DO and NO_3^- at deeper depths). However, the field study showed that denitrifying bacteria were more evenly distributed within the aquifer. Using the diffusion-limited model resulted in deeper movement of C and NO_3^- and microbial growth at deeper depths. Well data showed that DO decreased with depth but not DOC. This implied that the third model (aquifer matrix a source for C) was most appropriate. The model was most sensitive to the availability of organic C (maximum growth rate and microbial biomass were not sensitive parameters).

3.2.4 Simple Denitrification Models

Heinen (2003) reviewed over 55 N models to see how they simulated denitrification. Three different approaches have been used: 1) microbial growth models, 2) soil structure models, and 3) simplified process models. The microbial growth models simulate the dynamics of microbial populations. The soil structure models consider gaseous diffusion into and out of soil aggregates. These approaches were considered too complex for most modeling purposes and the review concentrated on the third category of simplified models.

Simplified models of denitrification take two forms. In one case, denitrification is described as a zero-order process (constant for fixed values of the reduction factors):

$$D_a = D_p f_N f_W f_T f_{pH} \tag{3.2.4-1}$$

where D_a is the actual denitrification rate [M M^{-1} T^{-1}, M L^{-3} T^{-1}, or M L^{-2} T^{-1}], D_p is the potential denitrification rate, and f_N, f_W, f_T, f_{pH} are dimensionless reduction factors for the soil NO_3^- concentration, water content, temperature, and pH. The potential denitrification rate is the rate expected under optimal conditions of high NO_3^- concentrations, saturation water contents, at a reference temperature, and normal soil pH.

In the second case, denitrification is described as a first-order process (increasing linearly with the concentration of NO_3^- for fixed values of the reduction factors):

$$D_a = k_d c_{NO3} f_W f_T f_{pH} \tag{3.2.4-2}$$

where k_d is the first-order rate constant [T^{-1}] and c_{NO3} is the concentration of NO_3^- in the soil. The same units for D_a and c_{NO3} are used in Equation 3.2.4-2 as are used for D_a and D_p in Equation 3.2.4-1.

When Equation 3.2.4-1 is used, the NO_3^- model reduction factor in most models reviewed by Heinen (2003) takes the form of a Michaelis-Menten equation:

$$f_N = \frac{c_{NO3}}{K_{MM} + c_{NO3}} \tag{3.2.4-3}$$

where K_{MM} is the Michaelis-Menten half-saturation constant (the value of c_{NO3} where f_N = 0.5) with the same units as c_{NO3}. In Figure 3-3, the effect of two different values of K_{MM} on the value of the NO_3^- reduction factor is shown.

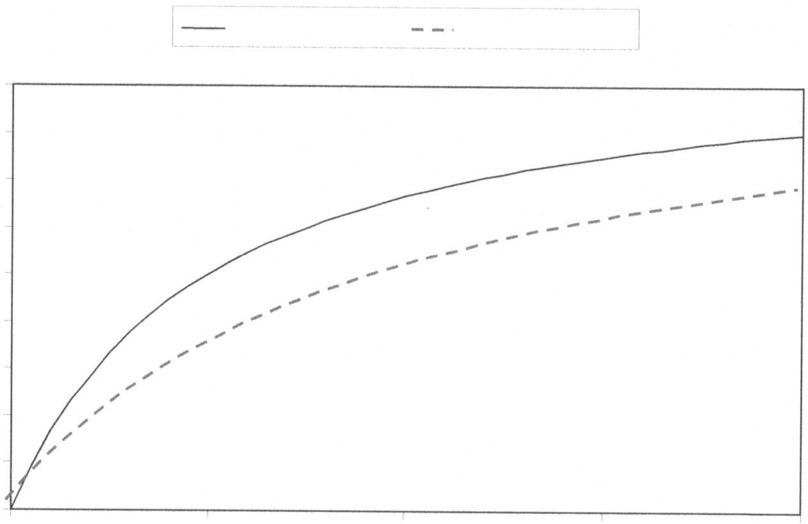

Figure 3-3. Nitrate-N Reduction Factor as a Function of NO_3^- Concentration for Two Values of K_{MM} Based on Equation 3.2.4-3.

For both Equation 3.2.4-1 and 3.2.4-2, many models use a power function for f_W:

$$f_W = \begin{cases} 0 & S < w_1 \\ \left(\dfrac{S - w_1}{w_0 - w_1}\right)^{w_2} & w_1 \le S \le w_0 \\ 1 & w_0 \le S \end{cases} \tag{3.2.4-4}$$

where S is the degree of saturation [-] defined as θ/θ_s, θ is the volumetric water content [L^3 L^{-3}], θ_s is the saturated volumetric water content [L^3 L^{-3}], w_0 is the value for S above which f_W = 1, w_1 is the value for S below which $f_W = 0$, and w_2 is the shape parameter that determines the steepness of the curve. Figure 3-4 shows the soil water reduction factor as a function of S based on Equation 3.2.4-4 for a soil with $w_0 = 1.0$, $w_1 = 0.6$, and $w_2 = 1.0$ used in the Crop Environment Resource Synthesis (CERES) wheat model (Jones and Kiniry, 1986).

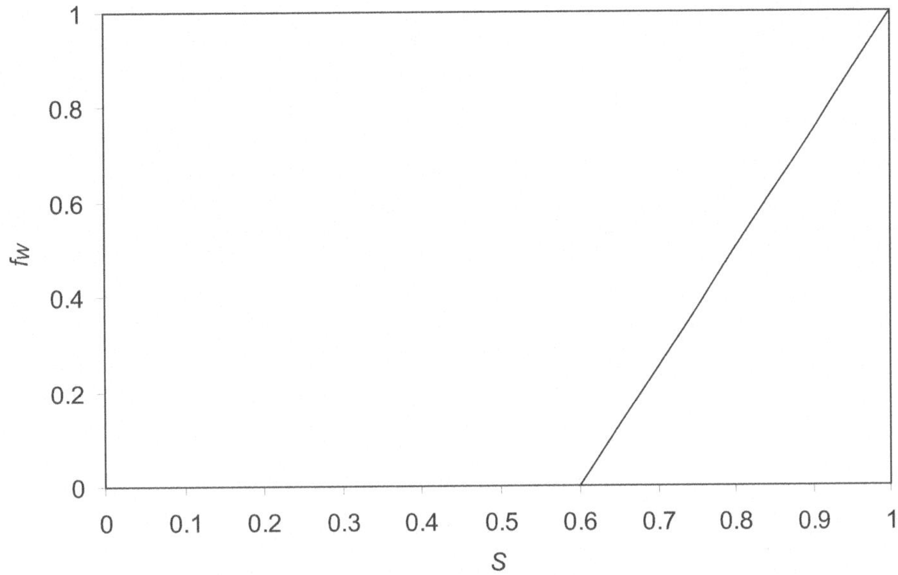

Figure 3-4. Water Reduction Factor as a Function of Degree of Saturation for Values of w_0 = 1, w_1 = 0.6, and w_2 = 1, Based on Equation 3.2.4-4.

Some models use an arctan function model that results in a sigmoidal shape for f_W instead of the linear model shown in Figure 3-4. Other models that compute denitrification losses on an annual basis use the average maximum water table level instead of S to determine f_W (Heinen, 2003).

The soil temperature reduction function is usually described in models using the van't Hoff equation or the Arrhenius equation (Heinen, 2003). Using the van't Hoff equation, f_T is:

$$f_T = \exp\left[k_H\left(T - T_{ref}\right)\right] = Q_{10}^{(T-T_{ref})/10} \qquad (3.2.4\text{-}5)$$

where k_H is the exponential increase coefficient [K^{-1}], T is the soil temperature [K], T_{ref} is the reference temperature [K] where $f_T = 1$, and Q_{10} is the increase in the reaction rate [-] expected for an increase in temperature of 10 °C or K. Typical values of Q_{10} are 2 to 3. The Arrhenius equation is similar:

$$f_T = \exp\left[k_A \frac{\left(T - T_{ref}\right)}{T_{ref}T}\right] \qquad (3.2.4\text{-}6)$$

where k_A is the exponential increase coefficient [K^{-1}]. The temperature reduction factor as a function of T for $Q_{10} = 3$ and $T_{ref} = 20$ is shown in Figure 3-5.

Optimal pH conditions for denitrification occur near pH 7.0-7.5 (Heinen, 2003). Many simple denitrification models do not include a reduction function for pH (f_{pH} in equation 3.2.4-1 or 3.2.4-2) probably because soil pH changes very little under normal agricultural conditions. When pH reduction functions are included in models, they usually consist of a bell-shaped curve with value of 1 around pH 7.0 and near zero for pH < 4 or pH > 10-11.

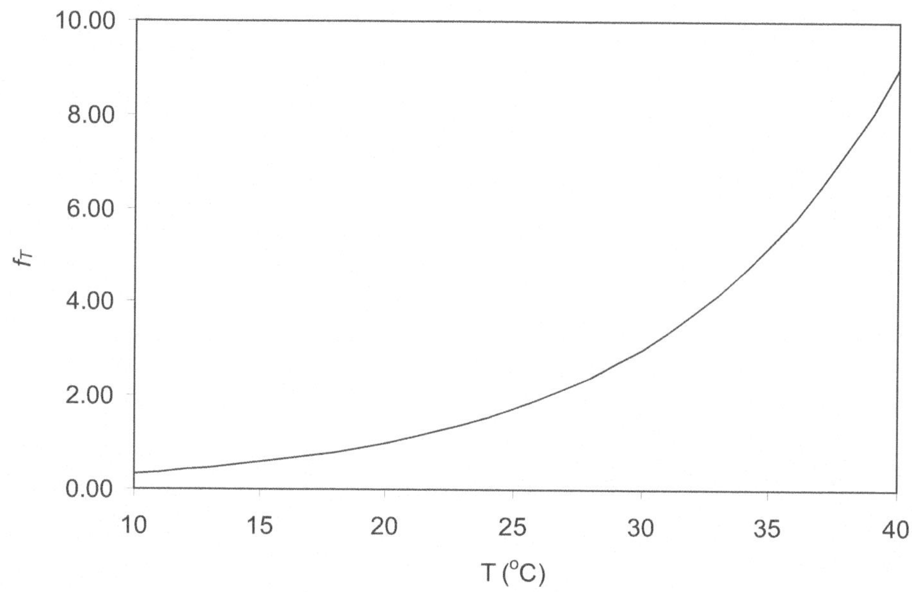

Figure 3-5. Temperature Reduction Factor as a Function of Soil Temperature for Values of Q_{10} = 3 and T_{ref} = 20 Based on Equation 3.2.4-5.

Heinen (2003) performed a sensitivity analysis for the simple denitrification models and found that they were most sensitive to the water reduction function, f_W in Equation 3.2.4-1 or 3.2.4-2. Heinen also compared the model predictions using Equation 3.2.4-1 to measurements from five different soils to see if the model parameter values could be differentiated by soil type through an optimization procedure. The data sets came from: 1) a loam soil overlying a clay, 2) a dry and wet sand, 3) a peat soil, 4) an anaerobic sandy peat soil, and 5) a heavy loam. Heinen found that the actual denitrification rate, D_a,was much less than the potential denitrification rate, D_p, in all cases indicating that the reduction functions were an important part of the simple model. Goodness of fit (R^2) for the different soils ranged from -0.37 to 0.90. There was no consistent pattern to the values for the model parameters found for the different soils so it was not possible to differentiate model parameters by soil type.

One of the simple denitrification models included in the review by Heinen (2003) was the NEMIS model developed by Henault and Germon (2000). The authors used a zero-order process to simulate denitrification (Equation 3.2.4-1) with reduction functions for NO_3^- concentration, soil water content, and temperature. They used extensive measurements of denitrification on intact cores from a site in France as a calibration dataset to determine the potential denitrification rate and the parameters for the water and NO_3^- reduction functions (water contents and NO_3^- concentrations of the cores were manipulated). The temperature reduction function was taken from the literature and consisted of a two-stage exponential growth model with a breakpoint at 11 °C. The model predictions were compared to field data from another study (Ryden, 1983) and to field data developed by the authors where water contents and temperature were measured continuously (Henault et al., 1998). The potential denitrification rates estimated for the calibration dataset and the two independent datasets were 3.94, 7.19, and 2.71 kg/ha/day. The model adequately predicted the dentrification rates of the two independent datasets (R^2 of 0.55-0.67 for the data from Ryden, 1983 and 0.90 for the data from Henault et al., 1998). A sensitivity analysis showed the model was most sensitive to uncertainties in the soil water content, especially the threshold water content where denitrification commences (w_1 in Equation 3.2.4-4). The authors suggested that this value could vary with soil texture and could be approximated by the field capacity water content. The model was also sensitive to the breakpoint in the two-stage temperature reduction function. Limits on denitrification due to C availability were not included directly in the model (perhaps due to the assumption that C is readily available in agricultural topsoils). Carbon limits could be included in the model, however, by assigning a value for potential denitrification rate under C-limited conditions.

Groffman and Tiedje (1991) suggested that simple models of denitrification should take soil texture and soil drainage class into account. In a field study, they found that the relationship between denitrification rates and air-filled porosity (one minus degree of saturation) varied among a loam, clay loam, and sand textured soil. In the sand, there was no relationship. In the loam soil, denitrification rates dropped sharply as air-filled porosity increased. In the clay loam soil, denitrification rates dropped more gradually as air-filled porosity increased. They attributed the pattern to differences in pore size distributions. The clay loam soil had smaller pores that remained anaerobic at higher air contents than the loam soil. In the sand, denitrification rates were uniformly low. They also found a relationship between respiration and the soil drainage class, but not soil texture. Other studies have shown that wet forest soils have high respiration rates at water contents near or at saturation. This is attributed to microbial populations that are adapted to low oxygen conditions. These populations can develop within weeks. Groffman and Tiedje (1991) found that respiration increased with air-filled porosity in well drained soils, decreased with air-filled porosity in poorly drained soils, and showed no relationship in somewhat poorly drained soils. This also suggested that wet soils, regardless of texture, may have microbial populations adapted to low oxygen conditions. The increased respiration rate in these soils could accelerate denitrification by using up available oxygen.

Studies have shown that denitrification rates in agricultural topsoils are highly variable in space with coefficients of variation approaching 100% (Folorunso and Rolston, 1984; Myrold, 1988; Parkin et al., 1987; Rice and Smith, 1982). Denitrifying bacteria require readily oxidizable OM, NO_3^- and a deficit of oxygen. Since oxygen and nitrate are relatively mobile compared to organic matter, denitrification hotspots of high oxidizable organic matter and/or sites where oxygen is restricted (by wetter conditions, for example) tend to develop. Parkin and Robinson (1989) developed a simple dentrification model that was unusual in that it was based on a stochastic approach. The denitrification rate was assumed to be proportional to the product of the denitrification enzyme activity (a measure of the potential denitrification rate) and the CO_2 production activity. A Monte Carlo approach was used to select values from lognormal distributions of enzyme activity and CO_2 activity. The model adequately predicted the mean and variance of measured denitrification rates in three validation data sets.

Soil aggregates containing OM could satisfy the conditions for a denitrification hotspot. Arah (1990) developed a denitrification model for soil aggregates (or nominal aggregates if a microsite is caused to be wet or contain high OM for some other reason). Movement of oxygen and nitrate into the interior of an aggregate was modeled as a diffusion process using the diffusion equation for spheres. It was a steady-state model with concentrations determined as a function of distance from the center of the sphere. A sink term in the equation represented the denitrification rate which was modeled with a Michaelis-Menten equation multiplied by a potential denitrification rate and a reduction function to account for the inhibitory effect of oxygen. The model showed that medium-size aggregates could have the greatest denitrification rate, depending on the diffusion coefficient used. Small aggregates did not become anaerobic because oxygen could diffuse to the center. Large aggregates became anaerobic but denitrification stopped when NO_3^- could not diffuse to the center. The model predicted concentrations of NO_3^- and oxygen as a function of radius within the aggregate. These values could be used to calculate the overall steady state rate of denitrification for an aggregate of a given radius. Arah (1990) also suggested a way to make the model stochastic by assuming a distribution of aggregate radii in the soil.

Bergstrom and Beauchamp (1993) developed a simple model of denitrification based on an equation similar to Equation 3.2.4-1. The reduction factors were developed from a "boundary analysis" of experimental data of denitrification rates as a function of the limiting factors (air-filled porosity, respiration rate, and mineralizable C). The data appeared as a scatter of points around a regression line indicating variability caused by measurement error and the interaction with other limiting factors. A line along the upper boundary of the data points represented the relationship between denitrification and the limiting factor unaffected by the other factors. The boundary line for air-filled porosity indicated that denitrification began at water contents near 50% of saturation. No evidence was found in the data that NO_3^- concentrations would be limiting in this model for agricultural topsoils. Model predictions were compared to measured denitrification rates in the field over time. The model described the time series reasonably well but failed to predict several denitrification events early in the season after a thaw.

3.2.5 HYDRUS Nitrogen Models

A number of studies have used HYDRUS-1D or HYDRUS-2D (Šimůnek et al.,2005) to model different forms of N, but only three studies have developed a HYDRUS N model for OWTS (Beggs et al., 2004; Hassan et al., 2008; Heatwole and McCray, 2007). The HYDRUS models, including the latest version which permits 3D simulations (Šimůnek et al., 2006), are capable of modeling a series of interacting solutes which can be adapted to forms of N. The HYDRUS models are finite element numerical models that simulate unsaturated and saturated water flow, temperature-dependent solute transport (including virus/bacteria), and heat transport.

Hanson et al. (2006) developed a HYDRUS-2D model for surface drip and buried line irrigation systems such as those used to irrigate vegetable crops in California. The model included N fertilizers added through the irrigation system. The fertilizer was assumed to be 50% urea-N, 25% NH_4^+-N, and 25% NO_3^--N. They modeled three forms of N that included hydrolysis of urea to NH_4^+ and mineralization of NH_4^+ to NO_3^-: urea $\rightarrow NH_4^+ \rightarrow NO_3^-$.

They did not include denitrification of NO_3^- because of a lack of information on denitrification rates and the dependency on temperature and water contents. Ammonium volatilization to NH_3 gas was not considered. The partial differential equation that described fate and transport of urea-N was a form of the CDE:

$$\frac{\partial \theta c_1}{\partial t} = \frac{\partial}{\partial x_i}\left(\theta D_{ij}\frac{\partial c_1}{\partial x_j}\right) - \frac{\partial q_i c_1}{\partial x_i} - \mu_{l,1}\theta c_1 \qquad (3.2.5\text{-}1)$$

where θ was the volumetric soil water content [$L^3\ L^{-3}$], c_1 was the concentration of urea-N in solution [$M\ L^{-3}$], t was time [T], x_i (i = 1,2) was the vertical and horizontal distance [L], D_{ij} was the dispersion coefficient tensor [$L^2\ T^{-1}$], and q_i was the water flux component [$L\ T^{-1}$]. The transformation of urea to NH_4^+ was assumed to be a first-order reaction and $\mu_{l,1}$ was the first-order rate constant [T^{-1}] for conversion in the dissolved form. The equation that described fate and transport of NH_4^+ was:

$$\frac{\partial \theta c_2}{\partial t} + \rho \frac{\partial s_2}{\partial t} = \frac{\partial}{\partial x_i}\left(\theta D_{ij} \frac{\partial c_2}{\partial x_j}\right) - \frac{\partial q_i c_2}{\partial x_i} - \mu_{l,2}\theta c_2 - \mu_{s,2}\rho s_2 + \mu_{l,1}\theta c_1 - s c_{r,2} \qquad (3.2.5\text{-}2)$$

where c_2 was the concentration of NH_4^+-N in solution [M L^{-3}], ρ was soil bulk density [M L^{-3}], s_2 was the concentration of NH_4^+-N adsorbed [M M^{-1}], and $\mu_{l,2}$ and $\mu_{s,2}$ were the first-order rate constants [T^{-1}] for conversion of NH_4^+ to NO_3^- in the dissolved form and adsorbed form, respectively. The next to last term in Equation 3.2.5-2 was a source term representing the mineralization of urea and was the same as the last term in Equation 3.2.5-1, except for the positive sign. The last term in Equation 3.2.5-2 was a sink term representing plant uptake of NH_4^+ where s was the water uptake rate [L^3 L^{-3} T^{-1}] and $c_{r,2}$ the concentration of NH_4^+ in the sink term [M L^{-3}]. The equation that described fate and transport of NO_3^--N was:

$$\frac{\partial \theta c_3}{\partial t} = \frac{\partial}{\partial x_i}\left(\theta D_{ij} \frac{\partial c_3}{\partial x_j}\right) - \frac{\partial q_i c_3}{\partial x_i} + \mu_{l,2}\theta c_2 + \mu_{s,2}\rho s_2 - s c_{r,3} \qquad (3.2.5\text{-}3)$$

where c_3 was the concentration of NO_3^--N in solution [M L^{-3}]. The last term in Equation 3.2.5-3 was a sink term representing plant uptake of NO_3^-. The second-to-last and next-to-last terms in Equation 3.2.5-3 were the same as in Equation 3.2.5-2 (except for the sign) and represented source terms due to conversion of dissolved and adsorbed NH_4^+ to NO_3^-. It was not clear from the paper if conversion of adsorbed NH_4^+ to NO_3^- was allowed to occur (if it did not occur then $\mu_{s,2} = 0$).

Adsorption of NH_4^+ was described by a linear equation:

$$s_2 = K_d c_2 \qquad (3.2.5\text{-}4)$$

where K_d was the distribution coefficient for NH_4^+ [L^3 M^{-1}]. It was assumed that there was no adsorption of urea or NO_3^-.

Hanson et al. (2006) ran their simulations for a loam texture soil with hydraulic properties taken from the HYDRUS soil database. They took their N model parameter values from the literature. For NH_4^+ adsorption, they assumed $K_d = 3.5$ cm^3/g and cited values from the literature of 3-4 cm^3/g (Lotse et al., 1992), 1.5 cm^3/g (Selim and Iskandar, 1992), and 3.5 cm^3/g (Ling and El-Kadi, 1998). For the first-order rate constant for conversion of urea to NH_4^+, they used $\mu_{l,1} = 0.38$/day and cited Ling and El-Kadi (1998) for using similar values of 0.36, 0.38, and 0.56/day. For the conversion of NH_4^+ to NO_3^-, they used $\mu_{l,2}$ (and perhaps $\mu_{s,2}$) = 0.2/day, citing similar values from the literature of 0.2 (Jansson and Karlberg, 2001), 0.02-0.5 (Lotse et al., 1992), 0.226-0.432 (Selim and Iskandar, 1992), 0.15-0.25 (Ling and El-Kadi, 1998), and 0.24-0.72/day (Misra et al., 1974).

The results from Hanson et al. (2006) showed very little build-up or movement of urea or NH_4^+ away from the source due to hydrolysis of urea and adsorption of NH_4^+. Nitrate moved out of the root zone and below a depth of 100 cm in both drip and buried line irrigation systems. Plant uptake and storage in the root zone were the largest components of the mass balance for N. Leaching losses accounted for 2-6% of the N balance.

Heatwole and McCray (2007) developed a model of N fate and transport below OWTS trenches using HYDRUS-1D. The objective was to assess the likelihood of N reaching groundwater at a depth of about 30 m at a site in Colorado (Todd Creek) where a large subdivision was proposed. A uniform and a layered profile were simulated. Ammonium sorption was not considered because it had been shown to be relatively unimportant compared to the N transformation processes. The authors used median first-order rate constants for nitrification (3.25 1/day) and denitrification (0.042 1/day) taken from the review of literature values by McCray et al. (2005). The top boundary of the model was the trench infiltration surface and the bottom was a free drainage boundary condition. Simulations were run until steady state was achieved (in less than 350 days). The results showed that NH_4^+ did not move beyond 30 cm below the trench. Using the median value for denitrification, NO_3^- concentrations in groundwater were well below the MCL for NO_3^- (10 mg/L). However, a sensitivity analysis showed that the model was most sensitive to the denitrification rate and there was a great deal of variation (3 orders of magnitude) in the literature values (McCray et al., 2005). Using denitrification rates from the lower end of the range resulted in NO_3^- concentrations exceeding the MCL at the depth of the water table.

Beggs et al. (2004) used HYDRUS-2D to simulate N fate and transport in a SDIS for onsite wastewater effluent. Beggs et al. (2004) used a model space that was rectangular with a width equal to the half space between drip emitters (30.5 cm) and a depth of 100 cm (soil surface at the top). The emitter was placed at a depth of 15 cm. Turfgrass was selected as the crop with the rooting depth to 30 cm. Rooting depth declined linearly from the surface to the maximum depth. Root uptake properties were taken from the HYDRUS database. Three soils textural classes (sandy loam, loam, and clay loam) were simulated with properties taken from the HYDRUS database, as well. Weather data were taken from long-term records for winter months (November through May) in California. It was assumed that there was no evaporation due to a full grass canopy. Potential transpiration daily average rates were taken from weather records. The N chain reaction models (Equations 3.2.5-1 to 3.2.5-3) were used to simulate ammonium, nitrate, and nitrite. Adsorption of ammonium was included and varied with soil textural class. In HYDRUS-2D, transformation rates for first-order reactions are constant. However, the authors used a pseudo-gas phase for ammonium and nitrate to make the reaction rates dependent on water content (nitrification optimal near field capacity and denitrification optimal at saturation). Carbon was not modeled so the effect of C and N transformations was not included. In addition to three soil textural classes, several different model scenarios were included. Nitrogen loss due to deep percolation varied from about 5 to 30% of the applied N. Losses were highest in the sandy loam soil and lowest in the clay loam soil. Plant uptake of water accounted for approximately 5-25% of the applied N. Denitrification losses were surprisingly high, varying from approximately 20-70%. Doubling the root zone depth had little effect due to ammonium adsorption and limited transpiration during the winter months. Once- daily pulses, as opposed to continuous application, accelerated nitrification and denitrification.

Hassan et al. (2008) also modeled an on-site wastewater SDIS. They used the two-dimensional version of HYDRUS and compared their model predictions to monitoring data for a site that serviced a restaurant in Virginia. The soil at the site had a low conductivity layer at a depth of 60 to 80 cm that favored lateral flow in the drainfield area where the slope range from 8 to12%. The OWTS included a pre-treatment unit and as a result the effluent N entering the drip system was all nitrate (ammonium or organic N). Equation 3.2.5-3 was used to simulate nitrate transport, plant uptake of nitrate, and denitrification. Denitrification was modeled as a first-order reaction with a rate constant of 0.002235 day^{-1} that did not vary with water content. The model space consisted of a sloping parallelogram, 20 m in length and 1.2 m in depth (soil surface at the top). Eight emitters were positioned 25 cm below the soil surface. The soil surface was modeled as an atmospheric boundary condition, the bottom surface was a free drainage boundary condition, and the sides were modeled as seepage faces. The monitoring data consisted of measurements of soil water pressure head and suction samplers for soil water nitrate. This data was used to estimate a nitrate mass balance for the drainfield system over the monitoring and 17-month modeling period. The largest mass balance component was storage of nitrate in the soil profile (48%), followed by plant uptake (27%), denitrification (22%), and nitrate losses and drainage (0.4%). There was excellent agreement between the model results (predicted water pressure heads and soil water nitrate concentrations).

Langergraber and Šimůnek (2005) developed a microbial growth model called Constructed Wetlands 2D (CW2D) to simulate N fate and transport in constructed wetlands. It consisted of a multi-component reactive transport module that ran in the HYDRUS model. The 13 components modeled in CW2D included:

- DO;
- three pools of OM: readily biodegradable OM, slowly biodegradable OM, and inert OM;
- four forms of N: NH_4^+, NO_2^- , NO_3^-, and N_2 gas;
- three groups of bacteria: heterotrophic bacteria responsible for hydrolysis, mineralization (aerobic growth), and denitrification (anoxic growth); autotrophic *Nitrosomonas* responsible for producing NO_2^-; and autotrophic *Nitrobacter* responsible for producing NO_3^-;
- inorganic P (IP); and
- a tracer of choice

Hydrolysis takes place independent of the oxygen condition. Aerobic growth of bacteria consumes DO, NH_4^+, and P. Anoxic growth of bacteria uses NO_3^- and NO_2^- and produces N_2 (denitrification). Adsorption is included for NH_4^+ and IP.

In CW2D, all transformations between components occur in the aqueous phase. The reaction rate for a given component (r_i, $i = 1, \ldots, 13$) [M L^{-3} T^{-1}] is given by:

$$r_i = \theta \sum_{j=1}^{R} v_j rc_j \tag{3.2.5-5}$$

where $j = 1, \ldots, R$ (R is the number of processes contributing to the production or loss of the component), υ_j is the stoichiometric factor for the reaction and component [M M^{-1}], and rc_j is the zero-order reaction rate for the component and process [M L^{-3} T^{-1}]. The zero order reaction rate for denitrification (rc_3) is:

$$rc_3 = \mu_{DN} \frac{K_{DN,O2}}{K_{DN,O2} + c_{O2}} \cdot \frac{c_{NO3}}{K_{DN,NO3} + c_{NO3}} \cdot \frac{K_{DN,NO2}}{K_{DN,NO2} + c_{NO2}} \cdot \frac{c_{CR}}{K_{DN,CR} + c_{CR}} \cdot \frac{c_{NH4}}{K_{DN,NH4} + c_{NO3}} \cdot \frac{c_{IP}}{K_{DN,IP} + c_{IP}} \cdot c_{XH} \tag{3.2.5-6}$$

where μ_{DN} is the maximum denitrification rate [T^{-1}]; $K_{DN,O2}$, $K_{DN,NO3}$, $K_{DN,NO2}$, $K_{DN,CR}$, $K_{DN,NH4}$, $K_{DN,IP}$, are the Monod equation saturation/inhibition coefficients [M L^{-3}] for DO (*O2*), NO_3^- (*NO3*), NO_2^- (*NO2*), readily biodegradable OM (*CR*), NH_4^+ (*NH4*), and *IP*, respectively; c_{O2}, c_{NO3}, c_{NO2}, c_{CR}, c_{NH4}, c_{IP}, are the concentrations [M L^{-3}] of DO, NO_3^-, NO_2^-, readily biodegradable OM, NH_4^+, and IP, respectively; and c_{XH} is the concentration of heterotrophic bacteria [M L^{-3}].

The exchange of oxygen from the gas phase to the aqueous phase is described by:

$$\frac{dc_{O2}}{dt} = ak_{aer,O2}\left(c_{O2,sat} - c_{O2}\right) \quad (3.2.5\text{-}7)$$

where $c_{O2,sat}$ is the saturation coefficient for DO [M L^{-3}], a is the soil air content [L^3 L^{-3}], and $k_{aer,O2}$ is the oxygen respiration rate [T^{-1}]. CW2D models heat transport as well and many of the parameters (e.g., $c_{O2,sat}$) are temperature dependent.

The CW2D model was developed to simulate a subsurface constructed wetland used in Austria where domestic wastewater infiltrates vertically through a sand bed. Langergraber and Šimůnek (2005) describe a single stage and double stage configuration of infiltration beds. In the single stage configuration, the bed was one m in width and 60 cm deep with a 10-cm layer of gravel at the bottom above a drain. The surface was an atmospheric boundary condition modified to allow ponding and the bottom was a seepage boundary condition. A daily loading rate of 40 L of wastewater was simulated, applied in 4 doses of 10 L at 6-hour intervals. The DO concentrations at depths of 5, 15 and 50 cm during a 24-hour period are shown in Figure 3-6. Concentrations decrease at 5 and 15 cm for about an hour with each dose but concentrations are relatively constant at a depth of 50 cm.

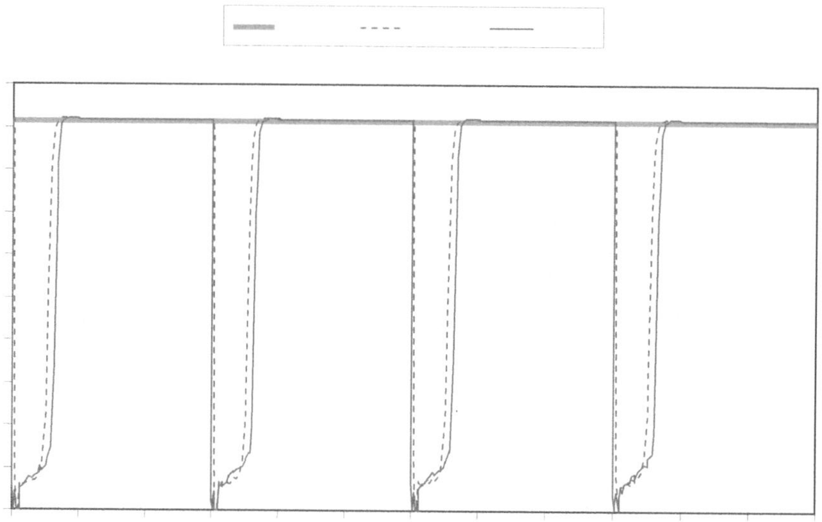

Figure 3-6. DO Concentrations at 5, 15, and 50 cm Depths Below the Infiltration Surface in a Single Stage Constructed Wetland with 4 Daily Doses Simulated by CW2D.

The concentrations of NO_3^- are shown for the same time interval and depths in Figure 3-7. Nitrate concentrations at 5 and 15 cm decrease with each dose as denitrification occurs under low DO conditions, but return to initial concentrations within an hour. Nitrate concentrations at 50 cm increase slowly over time.

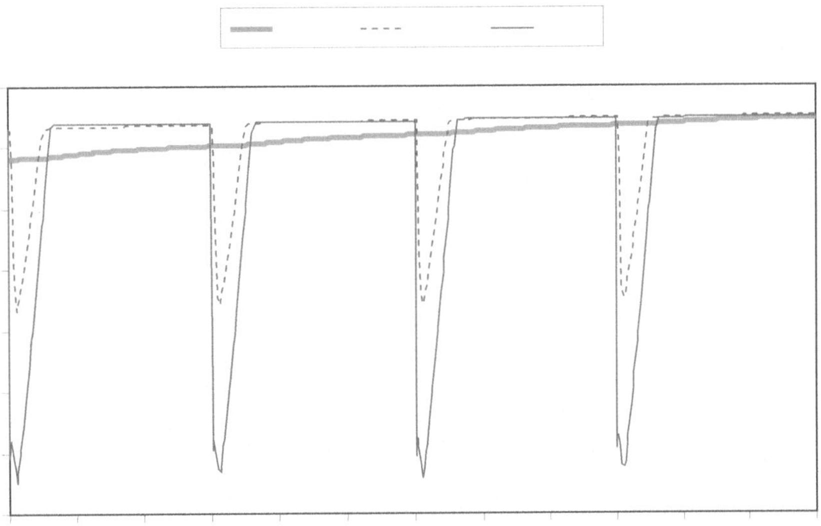

Figure 3-7. Nitrate Concentrations at 5, 15, and 50 cm Depths Below the Infiltration Surface in a Single-Stage Constructed Wetland with Four Daily Doses Simulated by CW2D.

In another study, Langergraber (2004) analyzed the role of plant uptake in wastewater assimilation. Plant uptake of nitrogen was relatively unimportant (1.9% of the influent nitrogen load) for a wetland treating municipal wastewater. For lower loaded systems such as gray water, plant uptake accounted for 46% of the nitrogen load.

3.2.6 SWAT Nitrogen Model

The Soil Water Assessment Tool (SWAT) is a widely used watershed-scale transport model that includes algorithms for modeling the different forms of soil N (Neitsch et al., 2002). SWAT simulates mineralization, nitrification, denitrification, and immobilization of N (Figure 3-8). A daily mass balance approach is used.

SWAT models five different pools of N (Figure 3-9) including NH_4^+, NO_3^-, and three forms of organic N (fresh, active, and stable) in each layer of soil.

Decomposition of the fresh organic N pool is allowed only in the soil surface layer. Mineralization from the active organic N pool to the NO_3^- pool is described by the equation:

$$N_{\min a,ly} = \beta_{\min}(\gamma_{tmp,ly}\gamma_{sw,ly})^{1/2}\, orgN_{act,ly} \qquad (3.2.6\text{-}1)$$

where $N_{mina,ly}$ is the amount of N mineralized (kg/ha N) in a soil layer during a time step, β_{min} is the rate coefficient for mineralization (unitless), $\gamma_{tmp,ly}$ is the nutrient cycling temperature factor (unitless), $\gamma_{sw,ly}$ is the nutrient cycling soil water factor (unitless), and $orgN_{act,ly}$ is the active organic pool N (kg/ha N). The default value in SWAT for β_{min} is 0.0003. Equation 3.2.6-1 is very similar to the source/sink terms in Equation 3.2.5-1, 3.2.5-2, and 3.2.5-3. It is modified by a geometric average of the nutrient cycling temperature and soil water factors.

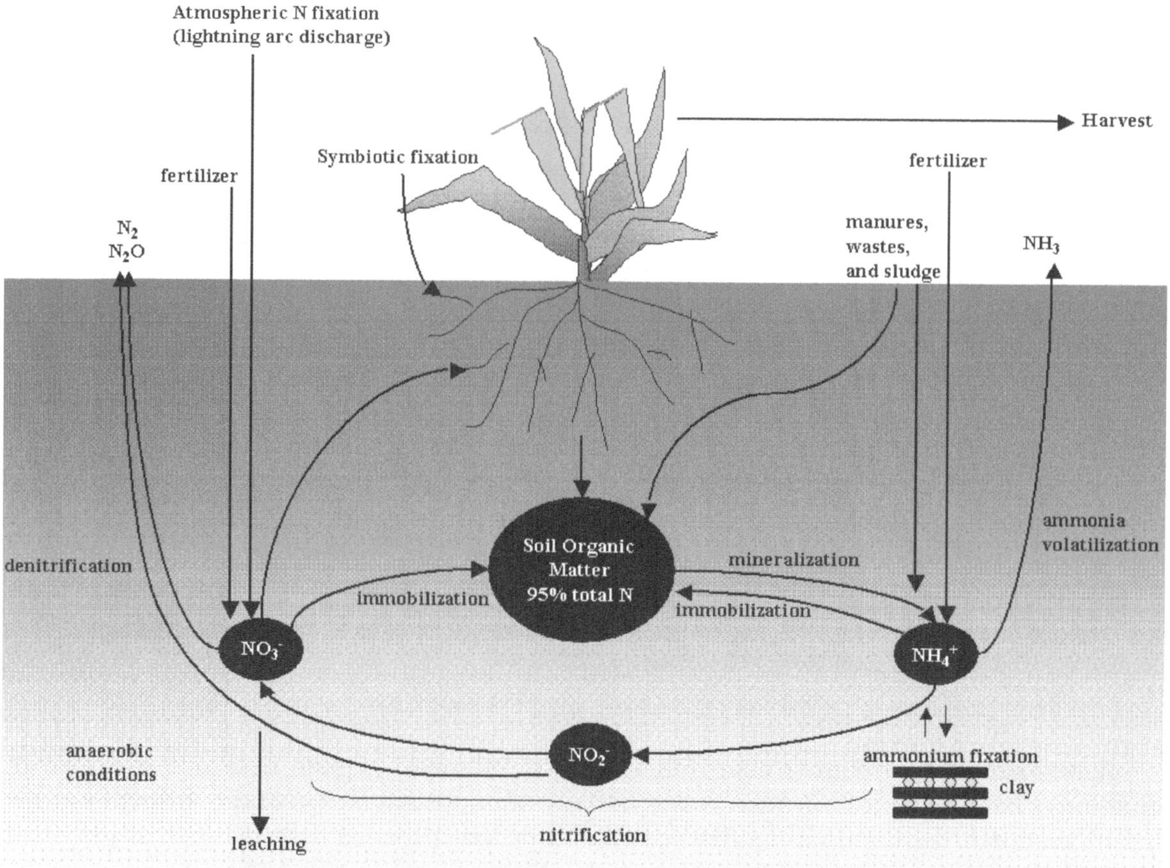

Figure 3-8. The N Cycle as Simulated by the SWAT Model.
Neitsch et al., 2002. Reprinted with permission.

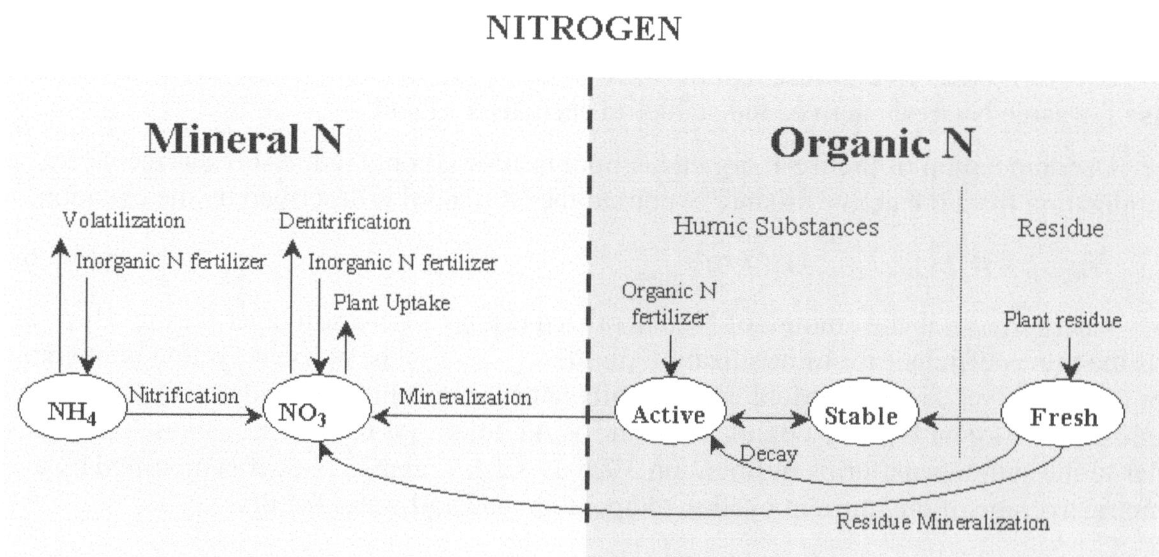

Figure 3-9. SWAT Soil N Pools and Processes that Move N in and out of Pools.
Neitsch et al., 2002. Reprinted with permission.

The nutrient cycling soil water factor is calculated as:

$$\gamma_{sw,ly} = \frac{SW_{ly}}{FC_{ly}} \qquad (3.2.6\text{-}2)$$

where SW_{ly} is the soil water content of the layer and FC_{ly} is the soil water content of the layer at field capacity (both in mm of water). Equation 3.2.6-2 shows that $\gamma_{sw,ly}$ will increase with water content in a linear fashion. The nutrient cycling temperature factor is calculated as:

$$\gamma_{tmp,ly} = 0.9 \frac{T_{soil,ly}}{T_{soil,ly} + \exp(9.93 - 0.312 T_{soil,ly})} + 0.1 \qquad (3.2.6\text{-}3)$$

where $T_{soil,ly}$ is the soil temperature of the layer (°C). A plot of $\gamma_{tmp,ly}$ as a function of $T_{soil,ly}$ is shown in Figure 3-10. The temperature factor increases nonlinearly with soil layer temperature and reaches a plateau at about 30 °C, as does the amount of fresh organic N mineralized.

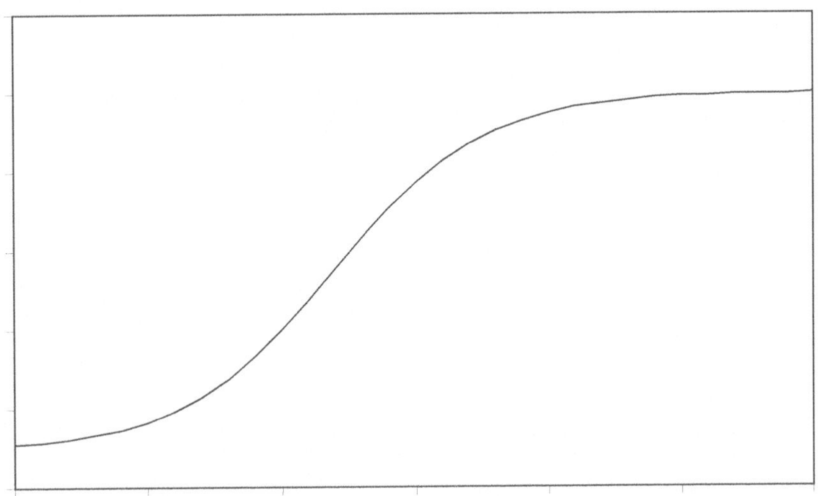

Figure 3-10. SWAT Nutrient Cycling Temperature Factor as a Function of Soil Layer Temperature.

SWAT includes volatilization of NH_4^+ but a depth factor reduces volatilization to zero for depths greater than 3 cm.

Nitrification is described by the following equation:

$$N_{nit} = NH4_{ly}\left[1 - \exp(-\eta_{nit,ly})\right] \qquad (3.2.6\text{-}4)$$

where N_{nit} is the amount of NH_4^+ converted to NO_3^- via nitrification in a soil layer during a time step (kg/ha N), $NH4_{ly}$ is the amount of NH_4^+ in the soil layer (kg/ha N), and $\eta_{nit,ly}$ is the nitrification regulator. The nitrification regulator is the product of the nitrification temperature factor ($\eta_{temp,ly}$) and the nitrification soil water factor ($\eta_{sw,ly}$) for the soil layer:

$$\eta_{nit,ly} = \eta_{tmp,ly}\eta_{sw,ly} \qquad (3.2.6\text{-}5)$$

The nitrification temperature factor is calculated as:

$$\eta_{tmp,ly} = 0.41\frac{T_{soil,ly}-5}{10} \qquad T_{soil,ly} > 5 \qquad (3.2.6\text{-}6)$$

The temperature factor is zero for soil temperatures below 5°C and increases linearly with temperature with a value of one at a temperature of about 30°C. The nitrification soil water factor is calculated as:

$$\eta_{sw,ly} = \frac{SW_{ly} - WP_{ly}}{0.25(FC_{ly} - WP_{ly})} \qquad (3.2.6\text{-}7)$$

where WP_{ly} is the soil layer water content at wilting point (mm of water). The soil water factor increases linearly with water content attaining a value of four at field capacity. The effect of soil water content on nitrification based on the above equations is shown in Figure 3-11 for a soil with 10 kg/ha NH_4^+-N in a soil layer. In this case, the temperature factor was held constant ($\eta_{tmp,ly}$ =1) with WP_{ly} = 10 mm and FC_{ly} = 30 mm. Nitrification increases rapidly from zero at wilting point to an asymptotic value near field capacity.

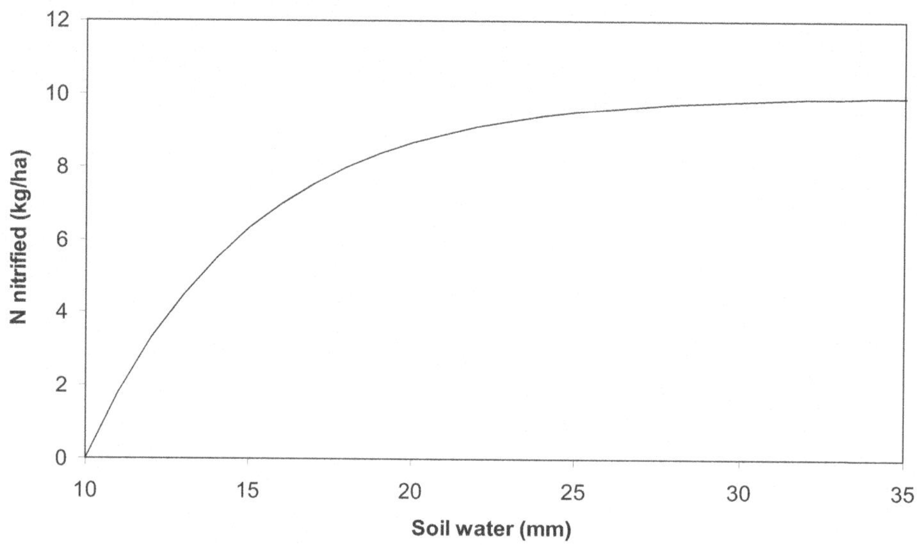

Figure 3-11. N Nitrified as a Function of Soil Water Content for a Soil with a Field Capacity Water Content of 30 mm in SWAT Based on Equations 3.2.6-4, 3.2.6-5, and 3.2.6-6.

SWAT models the amount of NO_3^- lost during a time step through denitrification ($N_{dent,ly}$ in kg N ha) using the following equation:

$$N_{dent,ly} = NO3_{ly}\left[1 - \exp(-1.4\gamma_{tmp,ly}\, orgC_{ly}\right] \qquad \gamma_{sw,ly} \geq 0.95 \qquad (3.2.6\text{-}8)$$

where $NO3_{ly}$ is the amount of NO_3^- in the soil layer (kg/ha N), $orgC_{ly}$ is the amount of organic carbon in the layer (%), $\gamma_{tmp,ly}$ is the nutrient cycling temperature factor (Equation 3.2.6-3) and $\gamma_{sw,ly}$ is the nutrient cycling soil water factor (Equation 3.2.6-2) for the soil layer. If $\gamma_{sw,ly}$ is less than 0.95, then there is no denitrifcation. This allows denitrification to occur for soil water contents slightly below field capacity up to saturation. The effect of soil temperature on denitrification based on Equation 3.2.6-8 is shown in Figure 3-12. Denitrification rates increase nonlinearly with temperature from 10 to 20°C and reach a plateau at about 25°C.

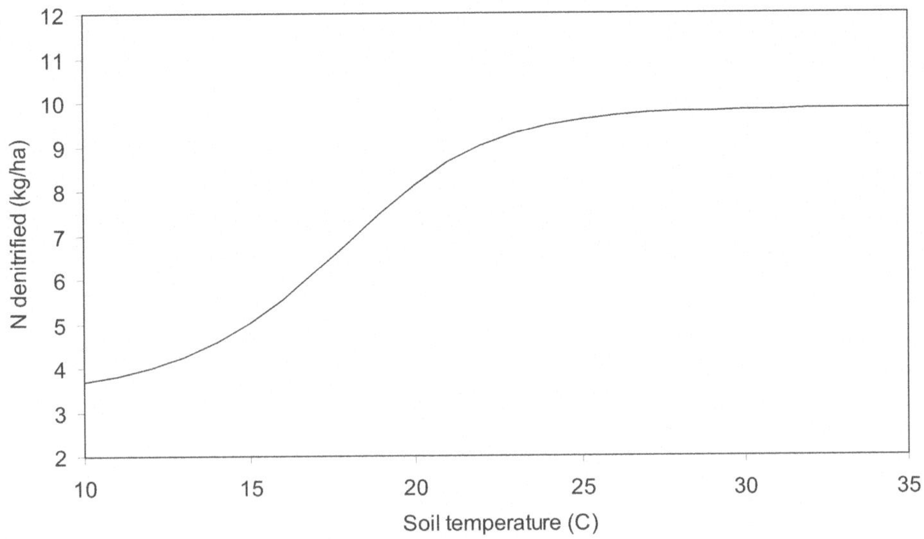

Figure 3-12. Nitrogen Denitrified as a Function of Soil Temperature in SWAT Based on Equation 3.2.6-8.

3.2.7 Riparian Zone Nitrogen Models

In this chapter, two riparian zone models are discussed. Riparian zones often represent ideal conditions for reducing NO_3^- losses. Many studies have examined the effect of riparian buffers on N fluxes to streams (see Lowrance et al., 1995 for a review). In general, these studies have shown that riparian buffers reduce NO_3^- fluxes in shallow groundwater flow to streams by 40-100%. The mechanism for removal of NO_3^- is primarily a combination of plant uptake and denitrification. Nitrate enters the riparian zone via shallow groundwater flow. Denitrification can occur in groundwater during the winter when the water table in the riparian zone is close to the soil surface where C levels and microbial populations are high. In the summer, however, most studies have shown that potential denitrification rates are very low at the depths where the water table occurs due to low C or microbial populations (Ambus and Lowrance, 1991; Groffman et al., 1992; Lowrance, 1992; Beare et al., 1994). During these months, plant uptake may temporarily store NO_3^- which is returned to the soil surface in leaf litter (Pinay et al., 1996). This is especially true in a mature forested riparian zone. The N is then mineralized and denitrified at the soil surface (Haycock et al., 1993; Hanson et al., 1994). Alternatively, roots may grow down into unsaturated soil in the summer and then when the water table rises in the winter, roots decay and become a C source, creating a hotspot for denitrification (Gold et al., 1998).

The Riparian Ecosystem Management Model (REMM) was developed to model N fate and transport in riparian buffer systems (Altier et al., 2002). It uses a mass-balance approach and operates on a daily time step. The riparian zone is modeled in two dimensions with the horizontal dimension being distance from the stream and the vertical distance being depth below the soil surface. A litter layer and three soil layers are included within the model space. Inputs to the model include surface runoff and lateral groundwater inflow from the upland area, as well as nutrient, sediment, and OM content in these flows.

REMM simulates three N pools within the soil and litter layers: residue, humus, and inorganic which includes NH_4^+ and NO_3^- (Figure 3-13). Although fixed ratios of C/N are shown in the figure for the humus pools, these ratios vary from 3 to15 for the active pool, 12 to 20 for the slow pool, and 7 to 10 for the passive pool, depending on the inorganic N availability.

Decomposition of the residue and humus pools results in addition to the NH_4^+ pool based on the C/N ratio of the organic pools, which varies from 150 in the structural residue pool to 8 in the active humus pool. Decomposition of OM follows the methods used in the Erosion-Productivity Impact Calculator (EPIC, Williams et al., 1983) and Nitrate Leaching and Economic Analysis Package (NLEAP) models (Shaffer et al., 1991). Decomposition is modeled as a first-order process multiplied by reduction factors for temperature, water content, and C/N ratio with factor values varying between 0 and 1. The temperature factor is an Arrhenius function with $Q_{10} = 2$. The water content factor for aerobic decomposition on day t ($WFac_{aerobic,t}$) [-] is based on degree of saturation (S as a %):

$$WFac_{aerobic,t} = \begin{bmatrix} 0.0075S & S \leq 20 \\ -0.253 + 0.0203S & 20 < S < 60 \\ 3.617\exp(-0.02274S) & 60 \leq S \end{bmatrix} \quad (3.2.7\text{-}1)$$

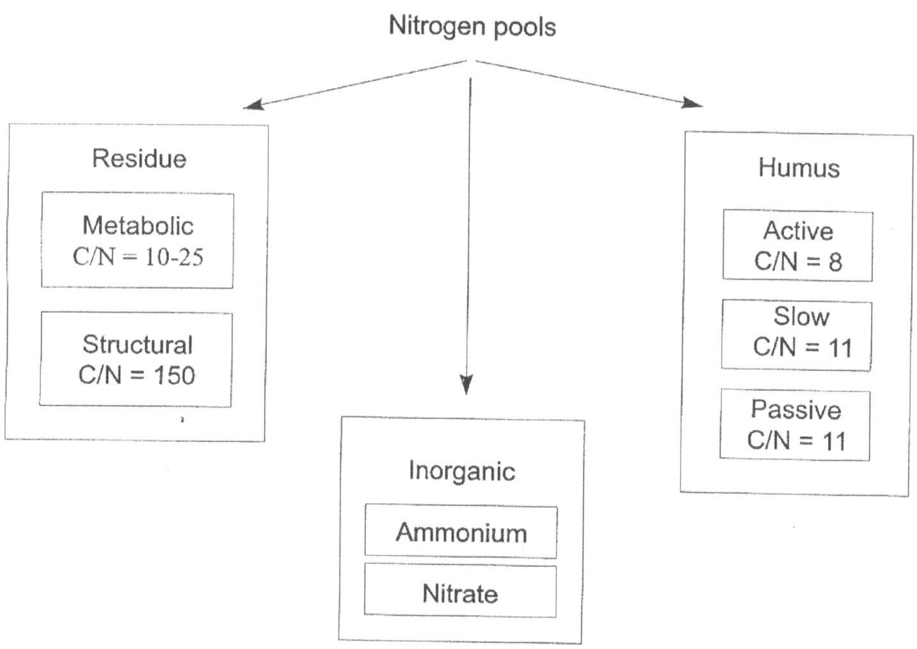

Figure 3-13. Nitrogen Pools Simulated in REMM (Altier et al., 2002)

The water content factor for aerobic decomposition (ammonification) is shown in Figure 3-14. Optimal decomposition rates occur when S is near 60% and rates decline above this water content.

The fluxes among N pools simulated in REMM are shown in Figure 3-15. Ammonium adsorption is simulated using a Freundlich adsorption equation. Nitrification is modeled as a first-order process using a Michaelis-Menten function for the first-order rate coefficient. Within the equation for the rate coefficient is an environmental limiting function which is the minimum of the aerobic water reduction function (Equation 3.2.7-1), the temperature reduction function, and a pH reduction function. The pH reduction function is optimal at pH between 7.0 and 7.4 and declines for pH outside this range.

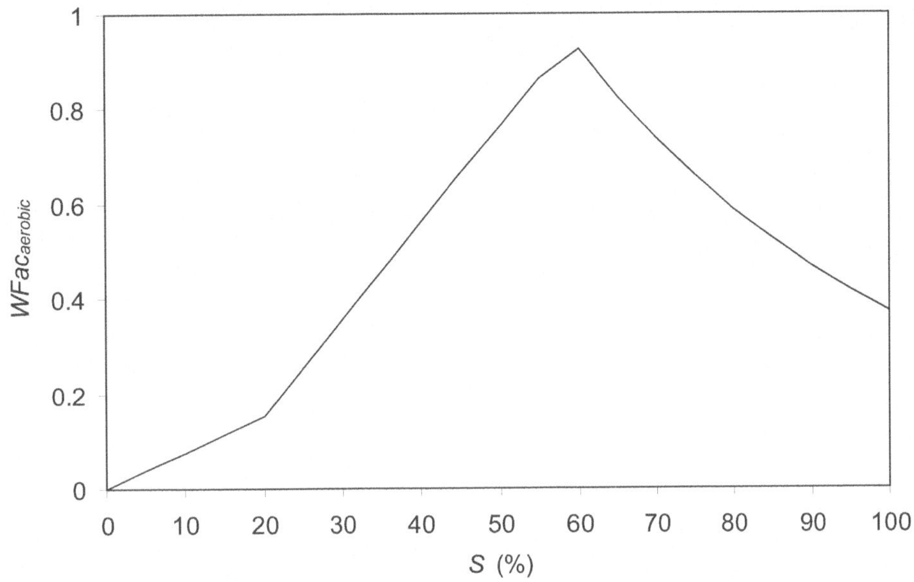

Figure 3-14. The Water Content Reduction Factor for Aerobic Decomposition (Ammonification) as a Function of Soil Layer Degree of Saturation in REMM.

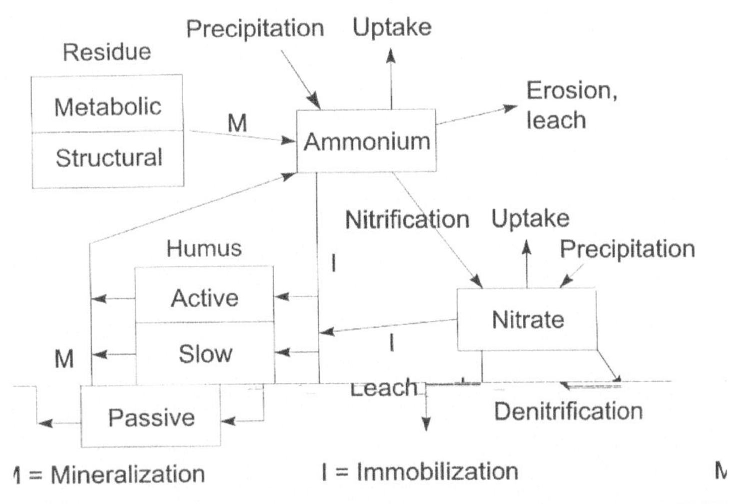

Figure 3-15. Fluxes Among N Pools in REMM (Altier et al., 2002).

Denitrification is assumed to be a zero-order process with respect to NO_3^- concentrations at concentrations above 2-5 mg/kg N, with the primary limitation being the availability of carbon. An Arrhenius temperature reduction function is included with $Q_{10} = 2$ and an optimum temperature of 35°C. An anaerobic reduction function is used to account for the effect of water content and carbon availability on denitrification. It includes a water reduction function for denitrification:

$$WFac_{anaerobic,t} = Min\begin{bmatrix} 1 \\ 0.000304\exp(0.0815S) \end{bmatrix} \quad (3.2.7\text{-}2)$$

The water reduction function for denitrification is shown in Figure 3-16. Denitrification is possible when the degree of saturation exceeds about 40%.

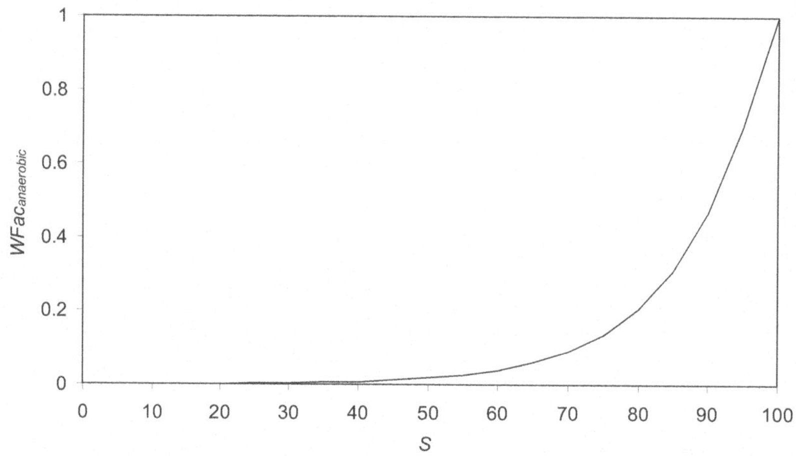

Figure 3-16. Water Reduction Function for Anaerobic Denitrification as a Function of Soil Layer Degree of Saturation in REMM.

The anaerobic reduction factor on day t ($AnaFac_t$) [-] is calculated as:

$$AnaFac_t = Min\left\{ \begin{matrix} 1 \\ Max\begin{bmatrix} 0.4 \\ XCoef \cdot AnaFac_{t-1} \end{bmatrix} \\ WFac_{anerobic,t} \cdot \left[2 - \exp(-CurvCoef \cdot CMin_{t-1})\right] \end{matrix} \right\} \quad (3.2.7\text{-}3)$$

where $XCoef$ is a coefficient [-] that limits the maximum amount by which $AnaFac_t$ can increase over the value for the previous day ($AnaFac_{t-1}$), usually a value of 1.5. $CurvCoef$ is a coefficient relating the amount of mineralizable C to its effect on anaerobiosis [ha/kg] and $CMin_{t-1}$ is the C mineralization on the previous day [kg/ha].

Another riparian zone model is the Riparian Nitrogen Model (RNM) developed in Australia (Rassam et al., 2005). It only considers N losses due to denitrification. The model assumes that the first-order potential denitrification rate exhibits an exponential decay with soil depth from the surface to the bottom of the root zone where it is zero. This is due entirely to the reduction in DOC with depth. Actual denitrification only occurs in groundwater so the intersection of the water table with the exponential decay depth function for potential denitrification rate determines the zone and rate of denitrification. The model simulates denitrification pulses during flooding events for ephemeral and low-order stream riparian zones. It also simulates denitrification that occurs in low-order stream riparian zones under base flow conditions where groundwater flows through the zone and intersects the root zone. The model determines the average denitrification rate and the time of duration for a given flooding event or base flow conditions. It then uses first-order kinetics to determine the N loss due to denitrification. Temperature effects are not taken into account.

3.2.8 Agricultural Nitrogen Models

Many comprehensive models of N fate and transport have been developed for agricultural conditions, in addition to the SWAT and REMM models. This review will describe several of these models as examples but not include all of the models.

A well-known agricultural N model is the Denitrification and Decomposition (DNDC) model developed by Li et al. (1992). The purpose of the model is to predict N_2O emissions from agricultural fields that occur as pulses after rainfall events. The model was developed as part of the scientific effort to assess greenhouse gas emissions, but it also predicts N_2 and CO_2 gas production. The model includes an organic N decomposition routine (ammonification and nitrification) and a denitrification (production of N_2O as well as N_2) routine. In between rainfall events, only the decomposition subroutine is active. When a rainfall event occurs, the denitrification subroutine is activated (and the decomposition subroutine inactivated). The denitrification subroutine executes on an hourly time step until the degree of saturation (S) in the top 20 cm of the soil profile falls below 0.40 or 240 hours have elapsed since the rain event started.

The soil profile (usually top 50 cm) is divided into layers (usually 2 cm in thickness). A thermal-hydraulic subroutine simulates water and heat flow between soil layers and each layer fills to saturation before the layer below receives water. Water flows out of the soil profile assuming a unit gradient. Evapotranspiration and crop uptake of water and nutrients are simulated. Soil physical properties (% clay, saturated hydraulic conductivity, water retention parameters, etc.) are assigned based on 12 textural classes including "organic" from Clapp and Hornberger (1978). Clay content also affects decomposition because clays absorb organic C and shelter it from decomposition.

In the DNDC model, soil temperature and soil moisture are key factors controlling both decomposition and denitrification. Reduction functions limit decomposition and nitrification at high and low temperatures and high and low S. Optimum conditions for nitrification occur at a temperature of 35°C and $S = 0.90$. Denitrification starts as soon as a soil layer becomes saturated. Denitrifier bacteria dynamics are simulated using multiple Michaelis-Menten terms that limit growth due to available carbon and nitrous oxides. The effect of soil pH is simulated such that at low pH (< 5) denitrification does not proceed beyond the production of N_2O. The denitrification rate increases exponentially with temperature reaching a maximum at 60-75°C.

Li et al. (1992) performed a sensitivity analysis of the DNDC model, using a combination of rainfall interval scenarios and a 1-year simulation with typical climatic data. They found that for all tests of the thermal-hydraulic subroutine, sandy soils were dryer than clayey or organic soils. The production of soluble C and NO_3^- were used as measures of decomposition. As the interval between rainfall events increased, NO_3^- (and to a much lesser extent, soluble C) concentrations increased. Decomposition was maximized when S was about 0.60. As soil temperature increased to 30°C, decomposition increased. The sensitivity analysis showed that N_2O and N_2 production increased as rainfall duration increased. Variations in annual precipitation had the greatest effect on annual total denitrification (increasing denitrification with increasing precipitation). Soil pH variation had the second largest effect on denitrification. Soil temperature also had a large effect. Changes in soil density and clay content had little effect (clay content was varied without changing soil hydraulic properties so that the effect on soil moisture conditions was not included). Clay content did have a large effect on N_2O production (as distinguished from the production of N_2O plus N_2). This was due to the stronger absorption by clays of N_2O than N_2.

Gabrielle et al. (2008) modified the CERES wheat model to include the production of N_2O from agricultural topsoils at three sites. Two different submodels were considered and tested against experimental data from fields in France. The submodels were based on two previously developed of N_2O emission: Nitrous Oxide Emission (NOE; Henault et al.,1998) and NGAS (Parton et al., 1996, 2001). NOE is a simple denitrification model based on the NEMIS model (discussed above in Section 3.2.4). NGAS is a more complex model of nitrification and denitrification that considers limitations due to C respiration rates, soil oxygen, labile C, NO_3^-, and soil pH. The field data showed NO_3^- and NH_4^+ concentrations were driven by fertilizer applications in the spring. The NOE submodel combined with the CERES model (CERES-NOE) did a better job of predicting N_2O emissions than the NGAS submodel combined with CERES (CERES-NGAS). The better performance of the NOE submodel may have been due to the fact that it required site-specific soil parameters such as the potential denitrification rate from sites tested in the study, whereas NGAS did not.

LEACHN is the N version of the LEACHM (Leaching Estimation and Chemistry Model) developed by Hutson and Wagenet (1992). It is a numerical model that simulates vertical movement of water and solutes within a soil profile. There are two versions of the LEACHM adapted for simulating N: LEACHNR uses a numerical solution to the Richards equation to simulate water flow and LEACHNA uses a capacity approach to fill each soil layer and includes a dual porosity system to simulate macropore flow. Jabro et al. (1995) compared predictions of monthly water flux and NO_3^- concentrations by both versions to measured concentrations in zero-tension pan lysimeters in a five-year field study that used different N fertilizer rates and a manure treatment. Model parameters for water flow and N transformations were calibrated using data from two of the five years. The calibrated N transformation parameters were within the range suggested in the model documentation. There was little difference between the two model versions and both monthly flow and N predictions were good for most months and treatments. However, both model versions failed to predict high NO_3^- losses in the high N fertilizer treatment in one of the verification years and low NO_3^- loses in two verification years of the zero fertilizer treatment. To get a good fit, it was necessary to adjust the N transformation parameters separately for each treatment.

DRAINMOD-N II is a comprehensive N model (Youssef et al., 2005) designed to run with the DRAINMOD hydraulic model (Skaggs,1978). These models are particularly suited to agricultural soils with artificial drainage. DRAINMOD-N II simulates plant uptake of N, mineralization/immobilization, nitrification, denitrification, volatilization, atmospheric deposition and fixation of N by legumes. The carbon cycle is also modeled based on routines in the CENTURY model (Metherell et al.,1993). Denitrification is modeled using Michaelis-Menten kinetics with rate reduction factors for relative soil water content, temperature, and soil depth (an exponential decay function limits denitrification at deeper depths due to depletion of carbon). In a case study simulation for a typical corn production system over 40 years, plant uptake accounted for 64% of the annual N use budget, denitrification 19%, and N loss through drain flow accounted for 14%. A sensitivity analysis showed that parameters controlling denitrification were the most sensitive (Wang et al., 2005).

Brief descriptions of several other well known agricultural N models are included below:

- ANIMO (Agricultural Nutrient Model) is a comprehensive nutrient cycling model developed in The Netherlands to simulate leaching of N from the soil surface to groundwater and surface waters (Kroes and Roelsma, 1997). It simulates ammonification, nitrification, denitrification, immobilization, and NH_4^+ adsorption as well as oxygen concentrations and temperature in the soil profile.

- CENTURY (Grassland and Agroecosystem Dynamics Model) is designed to simulate long-term soil OM dynamics in the topsoil in response to changes in management and climate (Metherell et al.,1993). The model uses a monthly time step and simulates fluxes of C, N, and P.
- EPIC is a mass-balance model designed to simulate the effect of soil erosion on crop production (Williams et al., 1983). The N routines are very similar to those in SWAT (described in Section 3.2.6 above) since SWAT was developed from EPIC.
- NLEAP is designed to estimate leaching of NO_3^- below the root zone (Shaffer et al., 1991). It includes ammonia volatilization, ammonification, nitrification, denitrification, and plant uptake.
- RZWQM (Root Zone Water Quality Model) is a comprehensive agricultural model that simulates N fate and transport using numerical solutions (Ahuja et al., 2000).

3.2.9 Conclusions

Many N models have been developed for agricultural applications but very few have been developed for OWTS and the processes that occur in the STU (Table 3-1). Since the most common forms of N in STE are organic N and NH_4^+, the N transformations of interest are ammonification, nitrification, and denitrification. Denitrification is highly sensitive to environmental factors, varying widely in space and time. The biggest question in modeling N in OWTS is under what conditions and to what extent does denitrification occur.

Two different approaches have been used for denitrification models: microbial growth models and simplified process models. The simple denitrification models are most sensitive to functions designed to account for the effect of soil water or air-filled porosity on denitrification (Heinen, 2003; Henault and Germon, 2000).

Several studies indicate that differences in soil texture, structure or drainage class are likely to affect denitrification, largely through their effect on soil water and oxygen availability (Groffman and Tiedje, 1991, Heinen, 2003; Henault and Germon, 2000; Li et al. (1992). However, studies have not shown a soil effect on N model parameters (Heinen, 2003).

Riparian zones often represent ideal conditions for reducing NO_3^- losses. Studies have shown that riparian buffers reduce NO_3^- fluxes in shallow groundwater flow to streams by 40-100%. This work and studies of agricultural topsoils indicate the importance of alternating aerobic and anaerobic periods and the effect of C in creating hotspots of denitrification.

A number of studies have used the HYDRUS framework (Šimůnek et al.,2005) to model different forms of N, but only one HYDRUS N model adapted for OWTS has been developed (by Heatwole and McCray, 2007). They found that their model was most sensitive to the denitrification rate (an input parameter) and there was a great deal of variation in the literature values. Beal et al. (2005), in a review of OWTS, suggested that models of sand filters might be adapted to OWTS. CW2D is a HYDRUS-based model of a sand filter that incorporates most of the features one might consider in a comprehensive microbial growth model, including a variable rate of denitrification due to changes in DO concentrations.

The HYDRUS 2D framework is especially suitable for modeling SDIS which may maximize N assimilation through plant update. It may be possible to expand the model domain to include shallow groundwater and base the design of an OWTS on the N loss at a down-gradient groundwater boundary such as the property boundary. This type of model domain could include important processes such as N transport in the water table capillary fringe (Abit et al., 2008). It is unrealistic to expect, however, that HYDRUS model domains can exceed more than about 10 m.

It may be possible to develop simple N models for STUs that will predict the effect of different soil types (texture, structure, and drainage class) by adapting the CW2D model to OWTS. This comprehensive model could then be run for the different soil types to generate N loss data. Simple regression models could then be developed from this data that relate soil types to predicted N losses.

Table 3-1. Summary of Models, Short Description and the N Process Simulated.

Reference	Model	Description	Nitrogen Processes
Ahuja et al., 2000	RZWQM	Comprehensive agricultural root zone model	Ammonification, nitrification, denitrification, plant uptake, and NH_4^+ sorption
Altier et al., 2002	REMM	Comprehensive riparian zone model	Ammonification, nitrification, denitrification, plant uptake, and NH_4^+ sorption
Anderson et al., 1998	OWTS	Groundwater transport numerical model	Dilution and dispersion
Arah, 1990	Agricultural topsoil model	Simple denitrification model for aggregates	Denitrification
Bergstrom and Beauchamp, 1993	Agricultural topsoil model	Simple denitrification model	Denitrification
Eichner et al., 1992	OWTS	Simple mass-balance model for predicting groundwater NO_3^- from OWTS in MA	Dilution
Frimpter et al., 1990	OWTS	Simple mass-balance model for predicting groundwater NO_3^- from OWTS in MA	Dilution
Gabrielle et al., 2008	CERES-NGAS and CERES-NOE	Crop growth model modified to predict N_2O emissions	Nitrification and denitrification
Hanson et al., 2006	HYDRUS-2D	N fertilizer movement in drip and buried irrigation systems	Ammonification and nitrification
Hantzsche and Finnemore, 1992	OWTS	Simple mass-balance model for predicting groundwater NO_3^- from OWTS in CA	Denitrification
Heatwole and McCray, 2007	HYDRUS-1D	N movement below a OWTS trench	Ammonification, nitrification, and denitrification
Henault and Germon, 2000	NEMIS	Simple denitrification model for agricultural topsoils	Denitrification
Hutson and Wagenet, 1992	LEACHN	Comprehensive agricultural topsoil model	Ammonification, nitrification, denitrification, plant uptake, and NH_4^+ sorption
Kinzelbach et al., 1991	Groundwater model	Groundwater multi-species model	Denitrification; NO_3^-, DOC, and DO transport; microbial growth
Kroes and Roelsma, 1997	ANIMO	Comprehensive agricultural topsoil model	Ammonification, nitrification, denitrification, immobilization, and NH_4^+ adsorption
Langergraber and Šimůnek, 2005	CW2D	Comprehensive numerical microbial growth model for constructed wetlands	Ammonification, nitrification, denitrification, plant uptake, and NH_4^+ sorption
Li et al., 1992	DNDC	Comprehensive mass-balance model for agricultural topsoils	Ammonification, nitrification, denitrification, and N_2O emissions
Metherell et al., 1993	CENTURY	Long-term OM dynamics model for agricultural topsoils	Ammonification, nitrification, denitrification, and plant uptake
Neitsch et al., 2002	SWAT	Watershed-scale, comprehensive model, mas- balance approach	Ammonification, nitrification, denitrification, ammonia volatilization, and immobilization
NJSOP, 1988	OWTS	Simple mass-balance model for predicting groundwater NO_3^- from OWTS in NJ	Dilution
Parkin and Robinson, 1989	Agricultural topsoil model	Simple denitrification model with stochastic C distribution	Denitrification
Rassam et al., 2005	RNM	Australia riparian zone model	Denitrification
Spiteri, et al., 2007	OWTS	Groundwater transport numerical model	Dilution, dispersion, and denitrification
Taylor, 2003	OWTS	Simple mass-balance model for predicting groundwater NO_3^- from OWTS	Denitrification and dilution
Weintraub et al., 2004	WARMF	Watershed-scale model modified to include effect of OWTS	Ammonification, nitrification, denitrification, and ammonia adsorption
Williams et al., 1983	EPIC	Agricultural topsoil erosion model	Ammonification, nitrification, denitrification, and ammonia adsorption
Youssef et al., 2005	DRAINMOD-N II	Comprehensive agricultural model for artificially drained soils	Ammonification, nitrification, denitrification, plant uptake, and ammonium sorption

3.3 Modeling Phosphorus in Onsite Wastewater Treatment Systems

Many watershed-scale models include estimates of P losses to surface water bodies, but they emphasize the overland flow path and use relatively simple approaches to model P losses through leaching (Nelson and Parsons, 2007). The focus of this review is on the leaching pathway of P loss. The framework for classifying models (Addiscott and Wagenet, 1985; Shaffer, 1995) that was discussed in Section 3.2 also applies to P models. A number of reviews of P models have been published. These include McCray et al. (2005), Radcliffe and Cabrera (2007), and Beal et al. (2005).

3.3.1 Phosphorus Transformations

The most common forms of P in STE are organic P and dissolved inorganic P (McCray et al., 2005). The P transformations of interest in modeling fate and transport of P within the STU are shown in Figure 3-17, and include:

- *mineralization*: the microbial transformation of organic P to dissolved inorganic and organic P (the reverse process is *immobilization*)
- *sorption*: the transformation of dissolved inorganic and organic P to sorbed P (the reverse process is *desorption*)
- *precipitation*:the transformation of dissolved inorganic P to Fe, Al, and Ca compounds (the reverse process is *dissolution*).

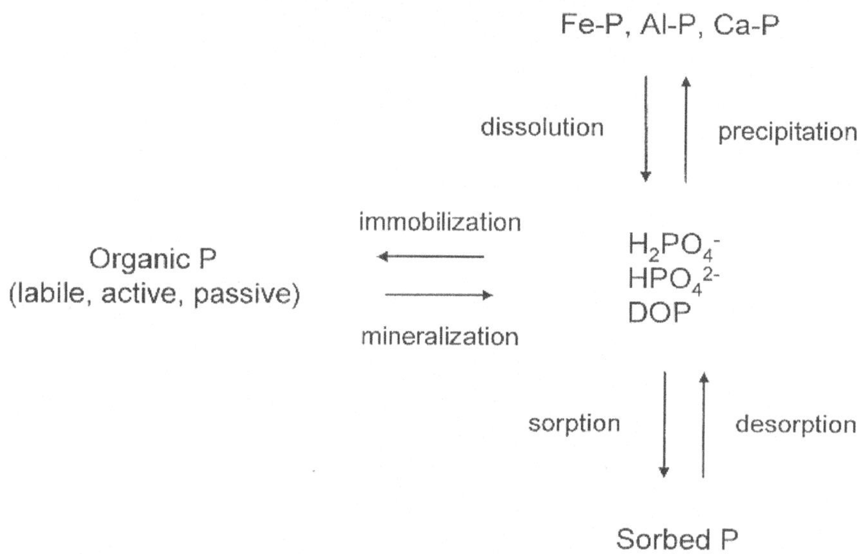

Figure 3-17. Phosphorus Transformations of Interest in the STU of OWTS.

Concentrations of P in soil solutions are very low compared to other nutrients in most cases (< 1 mg P/L) (Brady and Weil, 2008). Three forms of P are common in the solution phase: $H_2PO_4^-$, HPO_4^{2-}, and dissolved organic P (DOP). Plant roots can take up all of these forms of P. Soil pH affects the two anionic forms of P with $H_2PO_4^-$ being dominant at pH < 7 and HPO_4^{2-} being dominant at pH > 7. Dissolved organic P can be a large fraction of the solution P when wastes are added to soils. In the solid phase, organic P forms part of the labile (easily degraded), active, and passive pools of OM. Mineralization can occur under aerobic or anaerobic conditions so this transformation proceeds within the septic tank and most of the P in STE is in the form of dissolved inorganic P (McCray et al., 2005).

Forms of P in the solution phase are subject to sorption to clay minerals, OM, and iron and aluminum oxides. This sorption can be outer-sphere electrostatic attraction (weaker sorption) or inner-sphere substitution (stronger sorption). Soil pH affects sorption in that most of the P sorption sites are variable-charge sites with less sorption at higher pH. As a result, 1:1 clays such as kaolinite (high in variable charge sites) sorb more P than 2:1 clays such as montmorillonite. Sorption to OM is relatively weak. Adding OM to a soil can cause a reduction in sorption due to large organic molecules masking sorption sites, smaller organic molecules competing for sorption sites, or chelating of Fe and Al oxides to reduce potential sorption sites. In acid soils, Fe- and Al-P precipitates form and in alkaline soils Ca-P precipitates form.

Since soils differ in clay mineralogy; Fe, Al, and Ca content; and pH, they differ substantially in their capacity to retain P. Anaerobic conditions can cause microbial reduction of Fe oxides which releases P to solution. Laboratory studies have shown a two-stage retention process for soils and aquifers with an initial reversible reaction phase representing sorption, followed by a slow irreversible phase that may represent precipitation (Robertson, 1995). Phosphorus is much more strongly sorbed on iron oxides than calcite, the most common form of $CaCO_3$ (adsorption coefficient of 3000 cm^3/g for Fe oxide vs. 10 cm^3/g for calcite) (Spiteri et al., 2007).

Sorption of P can be modeled with a linear or nonlinear sorption isotherm, with or without kinetics. The simplest type of sorption is described by the linear non-kinetic (instantaneous) isotherm:

$$s = K_d c \qquad (3.3.1\text{-}1)$$

where s is the concentration of P sorbed [M M^{-1}], K_d is the distribution coefficient [L^3 M^{-1}] related to the strength of sorption, and c is the concentration of P in solution [M L^{-3}]. Nonlinear, non-kinetic sorption can be described using the Freundlich equation:

$$s = K_f c^{\beta} \qquad (3.3.1\text{-}2)$$

where K_f [$L^{3\beta}$ $M^{-\beta}$] and β [-] are Freundlich parameters. It may also be described using the Langmuir equation:

$$s = s_{max} \frac{K_l c}{1 + K_l c} \qquad (3.3.1\text{-}3)$$

where K_l is a Langmuir coefficient [L^3 M^{-1}] (similar to K_d) and s_{max} is the maximum amount of P that can be sorbed [M M^{-1}]. P sorption data for a sandy loam soil fit with the Freundlich and Langmuir isotherms are shown in Figure 3-18 from Nelson and Parsons (2007). The Langmuir equation shows a clear maximum sorption capacity but because of the difficulty in separating sorption from precipitation, P data often do not show a maximum.

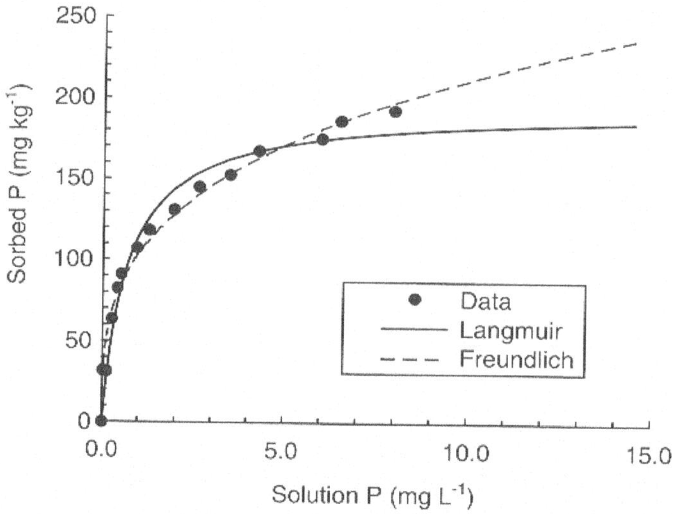

Figure 3-18. Freundlich and Langmuir Equations Fit to P Sorption Data for a Norfolk Sandy Loam.
Nelson and Parsons, 2007. Reprinted with permission.

The simplest kinetic sorption equation is that for linear sorption:

$$\frac{\partial s}{\partial t} = \alpha_k (K_d c - s) \qquad (3.3.1\text{-}4)$$

where t is time [T] and α_k is a first-order kinetic rate coefficient [T^{-1}].

The effect of linear, instantaneous sorption on transport of P can best be understood in terms of the retardation factor (R) [-]:

$$R = 1 + \frac{\rho_b K_d}{\theta} \qquad (3.3.1\text{-}5)$$

where ρ_b is the soil bulk density [M L^{-3}] and θ is the volumetric soil water content [L^3 L^{-3}]. The retardation factor is a ratio of the mean velocity of the soil or groundwater divided by the mean velocity of the sorbed constituent (P in this case).

McCray et al. (2005) reviewed the literature for model parameters suitable for simulating the fate and transport of P in OWTSs. They reported maximum P sorption capacities (s_{max}) that ranged from 0 to 17,600 with a median value of 237 mg/kg for 38 soils. For the 12 soils that were sand textures, the range in s_{max} was 15 to 1,368 with a median value of 40 mg/kg. For the thirteen soils that were sandy loam textures, the range in s_{max} was 116 to 1,640 with a median value of 405 mg/kg, clearly showing the importance of texture in P sorption. Texture of the other soils were not specified in the literature. McCray et al. (2005) also reported K_d values from the literature. The median value from 18 data sets was 15.1 cm^3/g. Using typical values for ρ_b and θ (1.5 g/cm^3 and 0.43 cm^3/cm^3, respectively), this indicates travel times for P that are on average 53 times as long as soil water or groundwater travel times.

Several questions are likely to be important in modeling P fate and transport in STUs. One is how sorption parameters such as K_d, s_{max}, and α_k vary among soils. Some studies have shown that P sorption isotherms for different soils can be collapsed into a single relationship if the sorbed P is expressed as a fraction of the total sorption capacity (Nelson and Parsons, 2007).

For sandy soils in Florida, Nair et al. (2004) calculated the sorbed P as the oxalate degree of P saturation (DPS_{Ox}) in percent:

$$DPS_{Ox} = \frac{Ox\text{-}P}{\alpha(Ox\text{-}Fe + Ox\text{-}Al)} 100 \qquad (3.3.1\text{-}6)$$

where *Ox-P*, *Ox-Fe*, and *Ox-Al* were the P, Fe, and Al extractable with 0.1 *M* oxalic acid + 0.175 *M* NH_4^+ oxalate and α was an empirical factor that varies with soil type (they used a value of 0.5 which is appropriate for sands). Since most of the sorption sites in these acid soils are due to Fe and Al, the sum of Ox-Fe and Ox-Al is a measure of the sorption capacity of the soil. When water-soluble P was plotted as a function of DPS_{Ox}, all of the samples fit a "split-line model" which increased in slope above a DPS_{Ox} breakpoint of about 20% (Figure 3-19). Above the breakpoint, more P is in solution because a significant portion of the sorption sites in these soils are filled indicating nonlinear sorption. For acid soils, it may be possible to estimate the isotherm parameters such as s_{max} and K_l in Equation 3.3.1-3 from the Fe and Al content of different soil textures and slopes of the split-line models.

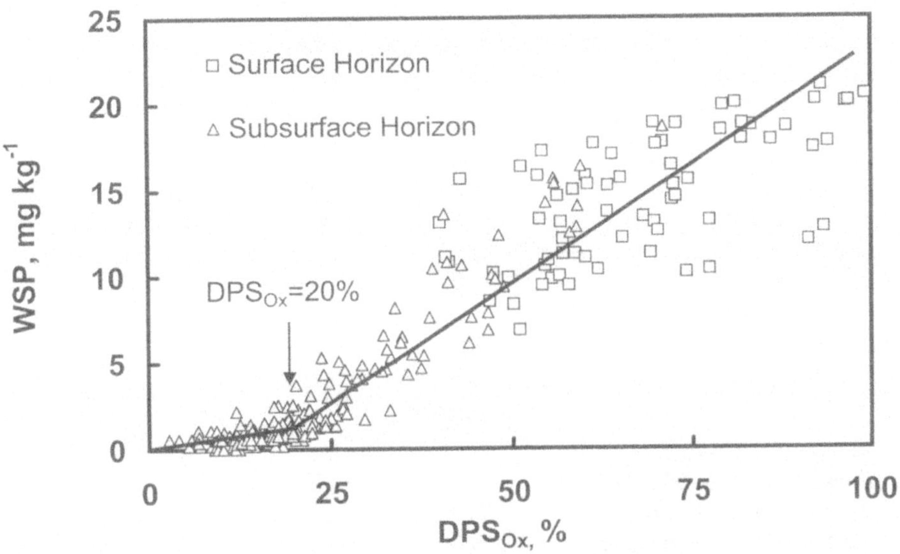

Figure 3-19. Relationship Between the Concentration of Water-Soluble P (WSP) and the Degree of P Saturation Calculated Using an Oxalate Extraction (DSP_{Ox}) for Surface and Subsurface Soil Samples.
Nair et al., 2004. Reprinted with permission from the Journal of Environmental Quality.

Another question is the extent of preferential flow. Soil structure can be expected to have an effect on P fate and transport because preferential flow will allow soluble P to bypass sorption sites. This effect can be incorporated into models, but assigning parameter values to different combinations of soil texture and structure will be difficult due to the lack of field data.

The last question is the degree to which sorption vs. precipitation occurs. The maximum sorption capacity may eventually be exceeded, especially in sandy soils or aquifers. If precipitation is not a significant loss, then concentrations in the soil solution and groundwater will eventually rise to a concentration close to that of the STE. If precipitation is significant, however, concentrations may rise to a value less than the effluent P concentration. Sorption can be modeled in a relatively simple manner as a source/sink term (as in Equations 3.3.1-1 through 3.3.1-4) in a CDE approach. To model precipitation with a CDE approach, a first-order decay term can be used (McCray et al., 2005). However, a multi-component reactive transport model that simulates fate and transport of the different forms of P and pH may be required. This approach for modeling P is less common, probably because of the difficulty in distinguishing between sorption and precipitation in field data and the added computational time and need for model parameter values in a multi-component approach.

3.3.2 Onsite Wastewater Treatment System Phosphorus Models

Lake loading models often assume that 100% of OWTS effluent P eventually reaches down-gradient waterbodies (Robertson, 1995). Other models assume a given failure rate as a percentage of the existing OWTSs within subwatersheds and then simulate a point source within each watershed that discharges P at concentrations typical of STE (Hummel et al., 2003). Very few models for fate and transport of P in OWTSs have been developed.

Weintraub et al. (2004) developed a modification to the WARMF model that simulated the fate and transport of P in soil beneath an OWTS. WARMF is a watershed-scale, mass-balance model used to predict nutrient loading to surface water bodies. The module simulated the development of a biozone or biomat (two cm in thickness) and the changes in hydraulic properties that occurred as the biozone matured. Sorption of P was included in the model. The module was calibrated using sand column data (Siegrist et al., 2002), adjusting the parameters for biozone hydraulic properties. The module was then tested in predicting the time to hydraulic failure of field-scale OWTS. The full model was applied to the Dillon Reservoir Watershed in Colorado where approximately 1,500 OWTS were present. Cumulative frequency distributions of STE flow and concentrations were used. The 50th percentile PO_4^{3-}-P concentration assumed was 9.8 mg/L. The model predicted that a scenario for converting from OWTS to a centralized sewer system would increase the total P load to surface waters. The OWTS P that was removed by soil sorption was greater than the treatment level that could be achieved by a wastewater treatment plant.

Lemonds and McCray (2003) used the SWAT model to simulate P loading from OWTS to the Dillon Reservoir, as well. This study is described in Section 3.3.3.

3.3.3 SWAT Phosphorus Model

The Soil Water Assessment Tool (SWAT) is a widely used watershed-scale transport model that includes algorithms for modeling different forms of soil P (Neitsch et al., 2002). Because of the difficulty in separating sorption from precipitation in field data, SWAT uses a simplified P cycle model developed by Jones et al. (1984) and Sharpley et al. (1984). This system has been adopted by a number of other models including EPIC (Williams, 1995), Groundwater Loading Effects of Agricultural Management Systems (GLEAMS; Knisel, 1993), Annualized Agricultural Nonpoint Source (AnnAGNPS; Bingner and Theurer, 2008) model, and Areal Nonpoint Source Watershed Environment Response Simulation 2000 (ANSWERS-2000; Bouraoui and Dillaha, 1996) model (Nelson and Parsons, 2007).

SWAT simulates mineralization, immobilization, and sorption/fixation of P (Figure 3-20). A daily mass-balance approach is used. SWAT models six different pools of P (Figure 3-21): three forms of mineral (inorganic) P and three forms of organic P in each layer of soil.

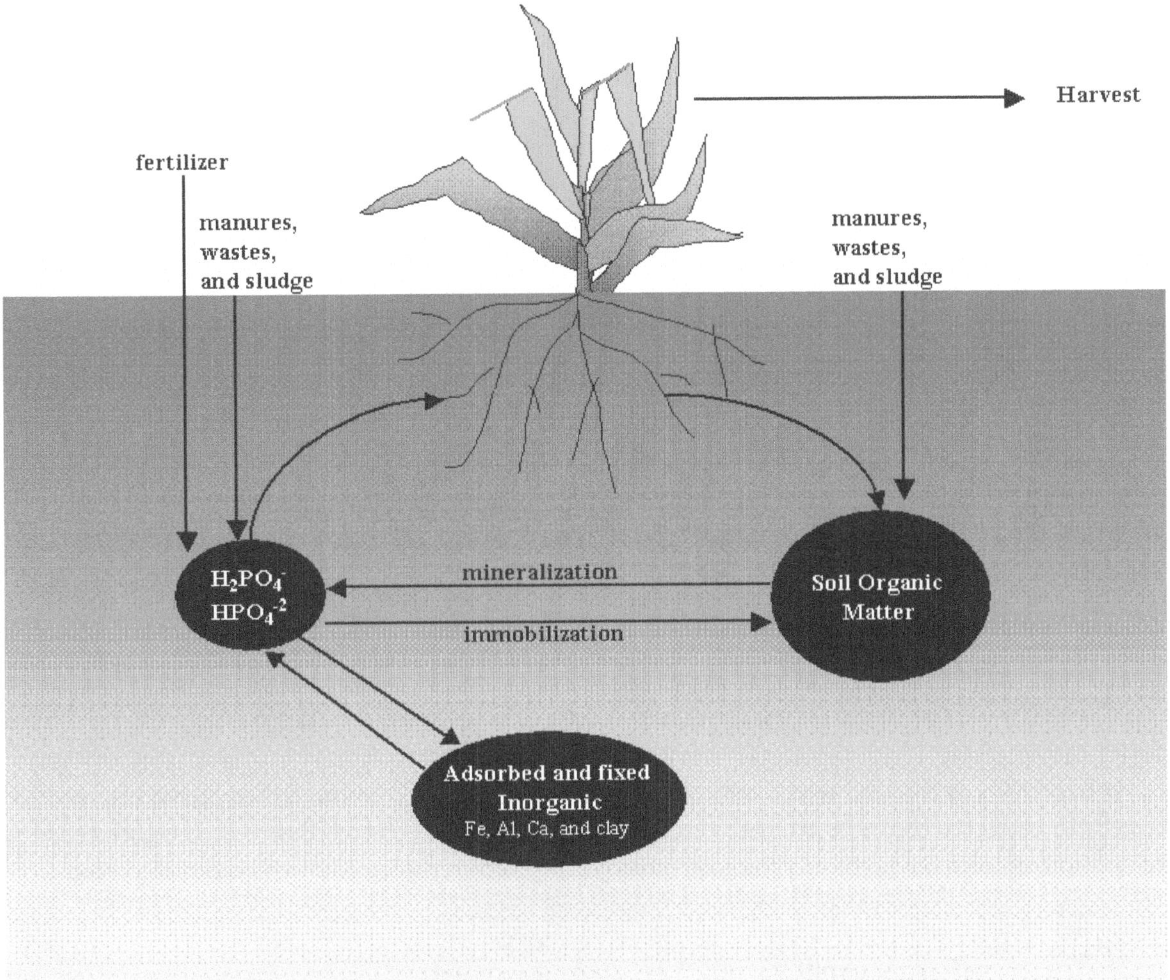

Figure 3-20. The P Cycle as Simulated by the SWAT Model.
Neitsch et al., 2002; Reprinted with permission.

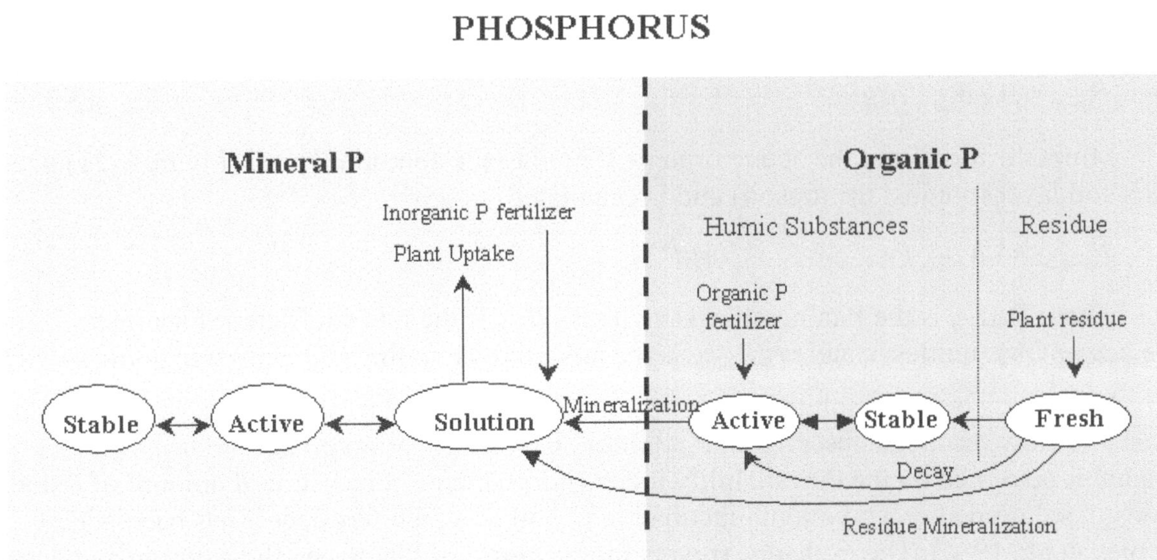

Figure 3-21. SWAT Soil P Pools and Processes that Move P in and out of Pools.
Neitsch et al., 2002; Reprinted with permission.

Decomposition of the fresh organic P pool is allowed only in the soil surface layer. Mineralization from the fresh organic P pool to the solution P pool (Figure 3-21) is described by the equation:

$$P_{minf,ly} = 0.8\,\delta_{ntr,ly}\, orgP_{frsh,ly} \qquad (3.3.3\text{-}1)$$

where $P_{minf,ly}$ is the amount of P mineralized (kg/ha P) during a time step, $\delta_{ntr,ly}$ is the residue decay rate constant for the soil layer (unitless) and $orgP_{frsh,ly}$ is the P in the fresh organic pool (kg/ha P). The residue decay rate constant is calculated:

$$\delta_{ntr,ly} = \beta_{rsd}\,\gamma_{ntr,ly}\,(\gamma_{tmp,ly}\,\gamma_{sw,ly})^{1/2} \qquad (3.3.3\text{-}2)$$

where β_{rsd} is the rate coefficient for mineralization of the fresh OM residue, $\gamma_{ntr,ly}$ is the nutrient cycling residue composition factor for the soil layer, $\gamma_{tmp,ly}$ is the nutrient cycling temperature factor, and $\gamma_{sw,ly}$ is the nutrient cycling soil water factor. The nutrient cycling residue composition factor ($\gamma_{ntr,ly}$) is calculated from an equation that takes into account the C:N ratio of the residue. All the parameters are unitless.

The nutrient cycling soil water factor is calculated as:

$$\gamma_{sw,ly} = \frac{SW_{ly}}{FC_{ly}} \qquad (3.3.3\text{-}3)$$

where SW_{ly} is the soil water content of the layer and FC_{ly} is the soil water content of the layer at field capacity (both in mm of water). Equation 3.3.3-3 shows that $\gamma_{sw,ly}$ will increase with water content in a linear fashion. The nutrient cycling temperature factor is calculated as:

$$\gamma_{tmp,ly} = 0.9\frac{T_{soil,ly}}{T_{soil,ly} + \exp(9.93 - 0.312T_{soil,ly})} + 0.1 \qquad (3.3.3\text{-}4)$$

where $T_{soil,ly}$ is the soil temperature of the layer (°C).

Mineralization from the fresh organic P pool to the active organic P pool (Figure 3-21) is described by the equation:

$$P_{dec,ly} = 0.2\,\delta_{ntr,ly}\, orgP_{frsh,ly} \qquad (3.3.3\text{-}5)$$

Mineralization from the active organic P pool to the solution P pool (Figure 3-21) occurs in each soil layer (not just the topsoil) and is calculated:

$$P_{mina,ly} = 1.4\,\beta_{min}\,(\gamma_{tmp,ly}\,\gamma_{sw,ly})^{1/2}\, orgP_{act,ly} \qquad (3.3.3\text{-}6)$$

where $P_{mina,ly}$ is the P mineralized (kg/ha P), β_{min} is the rate coefficient for mineralization of the active OM (unitless), and $orgP_{act,ly}$ is the amount of P in the active organic pool (kg/ha P).

Exchanges between solution, active, and stable mineral P pools (Figure 3-21) represent the fast and slow reactions observed in P retention in soils (sorption/precipitation). This exchange is governed by the P availability index (*pai*) which is a measure of amount of P that remains in solution after a six-month incubation of soil to which fertilizer P has been added (Sharpley et al., 1984). The exchange (reversible sorption) of P between the solution and active mineral pools (Figure 3-21) is controlled by a set of equilibrium equations:

$$P_{sol|act,ly} = P_{sol,ly} - minP_{act,ly}\left(\frac{pai}{1-pai}\right) \quad if \quad P_{sol,ly} > \min P_{act,ly}\left(\frac{pai}{1-pai}\right)$$
$$= 0.1\left[P_{sol,ly} - minP_{act,ly}\left(\frac{pai}{1-pai}\right)\right] \quad if \quad P_{sol,ly} < \min P_{act,ly}\left(\frac{pai}{1-pai}\right) \qquad (3.3.3\text{-}7)$$

where $P_{sol|act,ly}$ is the amount of P transferred between the soluble and active mineral pools in a layer, $P_{sol,ly}$ is the amount of P in the soluble mineral pool, and $minP_{act,ly}$ is the amount of P in the active mineral pool, all in kg/ha P. When $P_{sol|act,ly}$ is positive, P is being transferred from the solution to the active pool (sorption). When it is negative, P is being transferred from the active to the solution pool (desorption). Equation 3.3.3-7 shows that desorption is assumed to occur at one-tenth the rate of sorption.

Exchange between the active and stable mineral pools is governed by the following equations:

$$P_{act|sta,ly} = \beta_{eqP}(4\,minP_{act,ly} - minP_{sta,ly}) \quad if \quad minP_{sta,ly} < 4\,minP_{act,ly}$$
$$= 0.1\,\beta_{eqP}(4\,minP_{act,ly} - minP_{sta,ly}) \quad if \quad minP_{sta,ly} > 4\,minP_{act,ly} \qquad (3.3.3\text{-}8)$$

where $P_{act|sta,ly}$ is the amount of P transferred between the active and stable mineral pools in a layer, $minP_{act,ly}$ is the amount of P in the active mineral pool, and $minP_{sta,ly}$ is the amount of P in the stable mineral pool, all in kg/ha P. A slow rate of exchange is ensured by the equilibrium rate constant, $\beta_{eqP} = 0.0006$ for a daily time step. When $P_{act|sta,ly}$ is positive, P is being transferred from the active to the stable pool (which could represent precipitation). When it is negative, P is being transferred from the stable to the active pool (which could represent dissolution). Equation 3.3.3-8 shows that losses from the stable pool occur at one-tenth the rate of gains. It also shows that the stable mineral pool at equilibrium is four times larger than the active mineral pool.

Sharpley et al. (1984) provided equations for estimating *pai* (which is also called the "P sorption coefficient") given soil "labile P" concentration, clay content, and organic carbon content. Labile P is dissolved P plus a portion of weakly sorbed P. Sharpley et al. (1984) measured labile P as anion exchange resin extractable P (Sharpley, 2000). Labile P is a form of "bioavailable P" and there are other methods for measuring this fraction including the iron oxide strip method (Sharpley, 2000).

The soluble P lost in runoff is calculated:

$$P_{surf} = \frac{P_{sol,surf}\,Q_{surf}}{\rho_b\,depth_{surf}\,k_{d,surf}} \qquad (3.3.3\text{-}9)$$

where P_{surf} is the amount of P lost in surface runoff in a day (kg/ha P), $P_{sol,surf}$ is the amount of P in solution in the surface layer (kg/ha P), Q_{surf} is the amount of surface runoff in a given day (mm of water), ρ_b is the bulk density of the surface layer (Mg/m^3), $depth_{surf}$ is the thickness of the top layer (10 mm), and $k_{d,surf}$ is the P soil partitioning coefficient (m^3/Mg).

The P lost in eroded soil particles is calculated:

$$sedP_{surf} = 0.001 conc_{sedP}\frac{sed}{area_{hru}}\varepsilon_{P:sed} \qquad (3.3.3\text{-}10)$$

In calculating the amount of P that is leached, SWAT only considers P in the top 10 mm of soils as the source:

$$P_{perc} = \frac{P_{sol,surf}\, w_{perc,surf}}{10\rho_b depth_{surf} k_{d,perc}} \tag{3.3.3-11}$$

where P_{perc} is the amount of P moving from the surface layer (top 10 mm of soil) into the first soil layer (kg/ha P), $w_{perc,surf}$ is the amount of water percolating to the first soil layer from the surface layer on a given day (mm of water), and $k_{d,perc}$ is the ratio of the P concentration in the surface layer to the concentration of P in percolate (m^3/Mg).

Lemonds and McCray (2003) used the SWAT model to simulate P loading from OWTSs to the Dillon Reservoir in Colorado. No algorithms exist in SWAT for modeling OWTSs so the fertilizer management routine was used to simulate the effect of OWTSs. Fertilizer input parameters were adjusted to achieve the appropriate inorganic P input rate to the subsurface based on the number of OWTS in each subwatershed. A model sensitivity analysis showed that predictions of P loads were most sensitive to *pai* (Equation 3.3.3-7), $\varepsilon_{P:sed}$ (Equation 3.3.3-10), $k_{d,surf}$ (Equation 3.3.3-9), ρ_b (Equations 3.3.3-9 and 3.3.3-11), and the initial P concentration in the soil.

3.3.4 REMM Phosphorus Model

Riparian zones are not as effective in removing P as they are in removing N, but they still represent an important zone for trapping P (Lowrance et al., 1995). The Riparian Ecosystem Management Model (REMM) was developed to model nutrient fate and transport in riparian buffer systems (Altier et al., 2002). It is described in more detail in the chapter on modeling N (3.2). The modeling of P is very similar to SWAT (Section 3.3.3), but there are some interesting differences.

Like SWAT, REMM simulates inorganic (mineral) and organic (residue and humus) forms of P (Figure 3-22). Both models divide inorganic P into three pools. The labile P pool in REMM corresponds to the solution P pool in SWAT (Figure 3-21) and both models have an active and stable inorganic P pool. REMM divides residue P into two pools and humus P into three pools, unlike SWAT. REMM simulates dissolved forms of the inorganic-labile P pool and organic-active P pool (not shown in Figure 3-22). Adsorption and desorption between these pools is described using a Langmuir isotherm:

$$PConc_{Ads,t} = Q_0 \frac{b\, PConc_{Diss,t}}{1 + b\, PConc_{Diss,t}} \tag{3.3.4-1}$$

where $PConc_{Ads,t}$ is the adsorbed P concentration on day t (µg/g), $PConc_{Diss,t}$ is the dissolved P concentration (µg/L), and Q_0 (µg/g) and b (L/µg) are Langmuir coefficients corresponding to s_{max} and K_l in Equation 3.3.1-3). The effect of soil texture and pH on sorption coefficients is taken into account using the following equations:

$$Q_0 = -3.5 + 10.7\, clay + 49.5\, C \tag{3.3.4-2}$$

$$b = 0.061 + 1.7 \cdot 10^5 \cdot 10^{-pH} + 0.027\, clay + 0.76\, C \tag{3.3.4-3}$$

where *clay* is the percent clay, C is the percent carbon, and pH is the pH in the soil layer. These equations show that P sorption increases with clay and C content and decreases with pH.

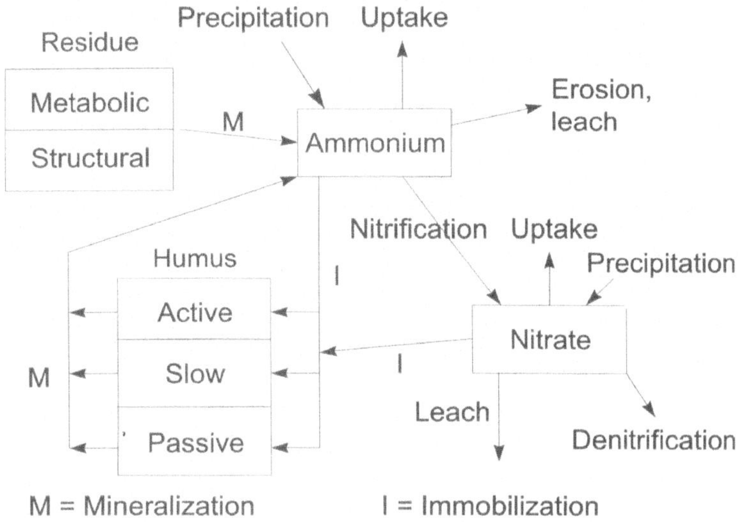

Figure 3-22. Phosphorus Pools and Transformations Simulated in REMM (Altier et al., 2002).

Exchange between the labile and active inorganic pools is governed by an equation similar to SWAT (Equation 3.3.3-7) but desorption and sorption occur at the same rate instead of desorption occurring at one tenth the sorption rate. Also, REMM includes temperature and water adjustment factors:

$$R_{la,t} = 0.1\left[P_{lb,t} - P_{act,t}\left(\frac{F_{lb}}{1-F_{lb}}\right)\right] TFac_{p,t}\, WFac_{p,t} \qquad (3.3.4\text{-}4)$$

where $R_{la,t}$ is the amount of P transferred from the labile to active inorganic pools on day t (kg/ha), $P_{lb,t}$ is the amount of P in the labile pool (kg/ha), $P_{act,t}$ is the amount of P in the active pool (kg/ha), and F_{lb} is the dimensionless P availability index (same as *pai* in Equation 3.3.3-7). The temperature factor is calculated:

$$TFac_{p,t} = \exp(0.115T_{soil} - 2.8) \qquad (3.3.4\text{-}5)$$

where T_{soil} is the soil temperature (°C). This function increases exponentially with temperature and has a value of unity at a temperature of about 25°C. The water factor is calculated:

$$WFac_{p,t} = \text{Minimum of}\begin{bmatrix} 1 \\ \dfrac{\theta_t}{\theta_{FC}} \end{bmatrix} \qquad (3.3.4\text{-}6)$$

where θ_t is the soil layer water content on day t (mm) and θ_{FC} is the soil layer water content at field capacity (mm). As such, the transfer in Equation 3.3.4-4 is accelerated as soil water content increases and reaches a maximum rate for water contents above field capacity.

Exchange between the active and stable inorganic pools is governed by an equation similar to SWAT (Equation 3.3.3-8), but desorption and sorption occur at the same rate:

$$R_{as,t} = K_{as}(4P_{act,t} - P_{stb,t}) \qquad (3.3.4\text{-}7)$$

where $R_{as,t}$ is the P transferred from active to stable forms on day t (kg/ha), $P_{act,t}$ is the amount of P in the active pool (kg/ha), $P_{stb,t}$ is the amount of P in the stable pool (kg/ha), and K_{as} is the first-order rate constant (1/day) corresponding to β_{eqP} in Equation 3.3.3-8. Instead of using a single value as in SWAT, K_{as} is adjusted for calcareous vs. noncalcareous soil:

$$K_{as} = \begin{bmatrix} 0.00076 & \text{calcareous soil} \\ \exp(-1.77F_{lb} - 7.05) & \text{noncalcareous soil} \end{bmatrix} \tag{3.3.4-8}$$

3.3.5 Convection Dispersion Equation Phosphorus Models

A number of models have been developed that use the CDE to simulate the transport of a single form of dissolved P and the associated sorbed/precipitated forms.

Hanson et al. (2006) developed a HYDRUS-2D model for surface drip and buried line irrigation systems such as those used to irrigate vegetable crops in California (described in more detail in Section 3.2 on modeling N). In the P transport model developed by Hanson et al., (2006), the partial differential equation that described fate and transport of P was a form of the CDE:

$$\frac{\partial \theta c}{\partial t} + \rho_b \frac{\partial s}{\partial t} = \frac{\partial}{\partial x_i}\left(\theta D_{ij} \frac{\partial c}{\partial x_j}\right) - \frac{\partial q_i c}{\partial x_i} - Sc_r \tag{3.3.5-1}$$

where c was the concentration of dissolved P in solution [M L^{-3}], x_i (i = 1,2) was the vertical and horizontal distance [L], D_{ij} was the dispersion coefficient tensor [L^2 T^{-1}], and q_i was the water flux component [L T^{-1}]. The last term in Equation 3.3.5-1 was a sink term representing plant uptake of dissolved P where S was the water uptake rate [L^3 L^{-3} T^{-1}] and c_r the concentration of dissolved P in the sink term [M L^{-3}]. Sorption of P was described by a linear isotherm (Equation 3.3.1-1) to replace $\partial s/\partial t$ in Equation 3.3.5-1. The authors noted that there is relatively little information in the literature on K_d values for P. They used a value derived from Silverbush and Barber (1983) of 59.3 cm^3/g. Kadlec and Knight (1996) reported a value of 19 cm^3/g and Grosse et al. (1999) reported a value of 185 cm^3/g. The results from Hanson et al., (2006) showed that there was very little movement of P away from the drip or line source due to sorption. They noted that their model did not include preferential flow and that this could have an important effect on the movement of P.

LEACHM was described in Section 3.2.8. The version of the model that simulates fate and transport of nutrients (N and P) is LEACHN. The labile and bound P pools represent inorganic P. The labile P pool includes a sorbed pool and a solution pool, similar to REMM and unlike SWAT. The bound pool can be considered precipitate or strongly sorbed P. The model can use Freundlich (Equation 3.3.1-2) or Langmuir (Equation 3.3.1-3) sorption isotherms for P sorption and desorption in the sorbed labile and bound P pools. Sorption sites can be included that are kinetic but sorption on these sites must be linear (Equation 3.3.1-4). At a given time step, the total concentration is compared to the maximum concentration that could occur if the solution phase was saturated with P and the sorbed phase was in equilibrium with this solution concentration according to the sorption isotherm. If the total P exceeds this maximum, then the excess P is considered precipitate in the bound pool.

Mansell et al. (1991) developed a CDE-based P transport model for 1D flow in soils that included two types of sites for kinetic sorption of inorganic P (slow and fast sorption sites). Rates of P sorption on fast sites were described by:

$$\frac{\partial s_1}{\partial t} = \frac{\theta}{\rho_b} k_1 c^m - k_2 s_1 \tag{3.3.5-2}$$

where s_1 was the concentration of sorbed P [M M^{-1}] on the fast sites, m was the Freundlich equation exponent (Equation 3.3.1-2) [-] ($m < 1$), and k_1 and k_2 were the forward (sorption) and backward (desorption) rate coefficients [units of T^{-1} when $m = 1$]. An identical equation described sorption on the slow sites with s_2 as the concentration of sorbed P and k_3 and k_4 as the forward and backward rate coefficients. Given sufficient time to equilibrate with a given soil solution concentration of P, the sites would reach an equilibrium described by a form of the Freundlich equation (Equation 3.3.1-2):

$$s = s_1 + s_2 = K_{eq} c^m \tag{3.3.5-3}$$

where: $$K_{eq} = K_1 + K_2 = \frac{k_1}{k_2} + \frac{k_3}{k_4} \tag{3.3.5-4}$$

Under these conditions (sufficient time to equilibrate with a given soil solution concentration, as might occur under very low soil water velocities), the retardation coefficient was a function of c:

$$R(c) = 1 + \frac{\rho_b}{\theta} K_{eq} m c^{m-1} \tag{3.3.5-5}$$

Equation 3.3.5-5 indicated that retardation would be high for low solution concentrations of P (for $m < 1$, R is inversely related to c) due to nonlinear sorption. Retardation increased as m increased (approaches unity or linear sorption which indicates no upper limit for sorption of P). Retardation also increased as the forward reaction coefficients increased relative to the backward reaction coefficients. Slow sorption kinetics (low values of k_1 and k_2) resulted in greater P mobility in that the soil solution did not have time to reach equilibrium with sorption sites. The model predictions were compared to soil column experiments using two sandy soils from Florida—the Immokalee and Myakka series. Four input concentrations of P were used. The two-site model adequately described the breakthrough curves in most cases, but the sorption parameter values had to be calibrated for each soil and input concentration. Differences between the soils were apparent in that slightly greater sorption and retardation occurred in the Myakka soil. The equilibrium $R(c)$ (appropriate for very low soil water velocities) for an input concentration of 5 mg/L was approximately 18 for the Immokalee soil and 21 for the Myakka soil. Retardation coefficients decreased for higher concentrations, indicating the limited extent of sorption due to a combination of kinetics and nonlinear sorption. At an input concentration of 100 mg/L, $R(c)$ was approximately 3 for the Immokalee soil. Using a model that assumed instantaneous sorption of P overestimated concentrations of P in the leachate on the rising limb of the breakthrough curve and underestimated tailing of P concentrations on the falling limb.

Chen et al. (1996) reported experiments to measure the sorption parameters in Equations 3.3.5-2 through 3.3.5-5 for the Myakka soil (95% sand). They observed that most batch sorption experiments use soil slurry mixtures with soil/solution ratios ($R_{s/w}$) that are much lower than the ratio that occurs in the field. In the field, the $R_{s/w} = \rho_b/\theta$, which for typical values of $\rho_b = 1.45$ g/cm^3 and $\theta = 0.45$ cm^3/cm^3 result in a $R_{s/w} = 3.2$ g/cm^3. To investigate the effect of soil/solution ratios, batch sorption experiments were performed at different $R_{s/w}$ (0.1 to 6.4 g/cm^3). Soil column leaching experiments were also performed using two column flow rates and four input concentrations of P. The experiments showed that the sorption parameters varied with $R_{s/w}$, with sorption decreasing at the higher soil/solution ratios typical of field conditions. The value of K_{eq} and m for a soil/solution ratio of 2.92 (which corresponded to the soil column experiments) were 67.6 cm^3/g and 0.378, respectively. The fraction of sites that were fast sorption (type-1) sites was found through calibration of the two-site sorption CDE model used in Mansell et al. (1991) and ranged between 15 and 37 % of the total sites (the remaining sites being slow or type-2 sites) for seven soil columns. Kinetic sorption experiments showed a rapid initial sorption phase followed by slower phase. More sorption occurred with low $R_{s/w}$. Increasing the input P concentration resulted in higher P mobility due to nonlinear sorption. Increasing the flow rate also resulted in higher P mobility due to sorption kinetics which limited sorption when residence times were short.

The HYDRUS model series (HYDRUS-1D, HYDRUS-2D, and the recently released HYDRUS which is capable of 2D and 3D simulations) provide a general framework for CDE-based modeling of P (Šimůnek et al., 2005, 2006). Sorption is described using the general equation:

$$s = \frac{kc^{\beta}}{1+\eta c^{\beta}} \qquad (3.3.5\text{-}6)$$

When $\beta = 1$ and $\eta = 0$, this is the linear sorption equation (Equation 3.3.1-1). When $\beta \neq 1$ and $\eta = 0$, this is the Freundlich equation (equation 3.3.1-2). When $\beta = 1$ and $\eta > 0$, this is the Langmuir equation (Equation 3.3.1-3). A two-site chemical non-equilibrium option is available that assumes instantaneous sorption on type-1 sites and kinetic sorption on type-2 sites (the parameter f is the fraction of total sites that are type-1 sites). Sorption kinetics on type-2 sites are described by:

$$\frac{\partial s}{\partial t} = \omega\left[(1-f)\frac{kc^{\beta}}{1+\eta c^{\beta}} - s\right] \qquad (3.3.5\text{-}7)$$

where ω is a first-order kinetic parameter [T^{-1}]. Equation 3.3.5-7 is similar to the kinetic sorption equation used by Mansell et al. (1991), Equation 3.3.5-2, described above.

Preferential flow (which may be very important for P transport in structured soils) can be simulated in the HYDRUS series of models in several ways. A physical non-equilibrium option is available that divides soil water into mobile (flowing) and immobile (stagnant) phases:

$$\theta = \theta_m + \theta_{im} \qquad (3.3.5\text{-}8)$$

where θ_m is the water content of the mobile phase [L^3 L^{-3}] and θ_{im} is the water content of the immobile phase [L^3 L^{-3}]. The exchange of solutes between the mobile and immobile phases is described by:

$$\left[\theta_{im} + \rho_b(1-f)\frac{k\beta c^{\beta-1}}{(1+\eta c^{\beta})^2}\right]\frac{\partial c_{im}}{\partial t} = \omega(c_m - c_{im}) \qquad (3.3.5\text{-}9)$$

where ω is the mass transfer coefficient [T^{-1}], c_m is the concentration in the mobile phase [M L^{-3}], c_{im} is the concentration in the immobile phase [M L^{-3}], and f is the fraction of the total water that is mobile. Preferential flow can be simulated by assuming that the mobile water phase represents macropore flow and the immobile phase represents matrix flow. Alternatively, the mobile water phase can represent inter-aggregate space and the immobile phase can represent intra-aggregate space. The sorption parameters can be different for the mobile and immobile phases to simulate bypass flow of sorption sites in the matrix. The physical non-equilibrium approach has been extended to water flow in recent modifications of the HYDRUS models in that water as well as solutes can flow between the mobile and immobile phases and the immobile phase is not required to be entirely stagnant (Šimůnek et al., 2005, 2006).

3.3.6 Multi-Component Transport Phosphorus Models

An example of a multi-component transport P model is the groundwater model developed by Spiteri et al. (2007). The authors used a 1D model that simulated horizontal transport of P in the groundwater and compared model results to data from two sites in Canada where OWTS had produced narrow groundwater plumes. About 15 years of transect well data was available at both sites. Fifteen interacting solutes were modeled including N, P, and DOC with 21 reaction pathways. They included two fractions of DOC: a labile fraction in the OWTS effluent and a less labile fraction in the natural groundwater. Three forms of dissolved P were simulated: $H_2PO_4^-$, HPO_4^{2-}, and PO_4^{3-} (favored at very high pH). Phosphorus reactions included release of P through mineralization of OM, fast reversible sorption on iron oxides or calcite, slow reversible sorption or diffusion into solid phases, and precipitation/dissolution of hydroxylapatite ($Ca_5(PO_4)_3OH$), vivianite ($Fe_3(PO_4)_2{\cdot}8H_2O$), and stringite ($FePO_4{\cdot}2H_2O$). A nonlinear (Langmuir) isotherm was used for fast sorption. An empirical equation for slow sorption was used that had an upper limit for sorption capacity (this distinguished it from precipitation which would not have an upper limit). Some of the reaction parameters came from the literature and others were fitted using the model predictions of well data concentrations. The fitted parameters included the slow P sorption constant.

The water table was two to three m below the OWTS infiltration beds at both sites described by Spiteri et al. (2007). The "Cambridge" site was a calcareous sandy aquifer (20% $CaCO_3$) and "Muskoka" site was a non-calcareous sand aquifer (0.5% Fe oxides). Both sites had long narrow plumes with very little evidence of dispersion or dilution (Robertson, 1995). Up to 90% of the effluent DOC was removed in the unsaturated zone between the infiltration bed and the water table (about 2 m) through oxic degradation, but the remaining DOC was still a significant recharge input. Oxygen and DOC concentrations decreased with distance in the plume according to the well data, indicating aerobic degradation of DOC. At the Cambridge site, the oxygen levels were depressed in the plume (0.03-0.06 μM) compared to the surrounding groundwater (0.13-0.25 μM).

Phosphorus retention (sorption and precipitation) in the unsaturated zone removed 23% of effluent P at the Cambridge (calcareous) site and 99% at the Muskoka (Fe-rich) site. At the Cambridge site, well data from 10, 14, and 17 years of operation showed an expanding plume of P but concentrations in the core of the plume were consistently about 3 mg/L (Robertson, 1995). Further attenuation within the plume was minimal despite the much longer flowpath compared to the unsaturated zone. Robertson (2003) stated that this was symptomatic of a precipitation reaction. If the retention mechanism in the unsaturated zone was sorption, then concentrations should have increased over time as sorption sites within the unsaturated zone were filled. The mean velocity of the groundwater at the Cambridge site was about 20 m/yr, whereas the P plume was advancing at a rate of about 1 m/yr, indicating a retardation factor of 20. At Cambridge, the pH of the plume was near neutral due to the buffering effect of $CaCO_3$, but at the Muskoka site the pH dropped to 4.9 in the plume. The drop in pH was due to nitrification and the absence of buffering capacity.

The model of Spiteri et al. (2007) was confined to transformations within the groundwater and did not include retention processes (sorption and precipitation) in the unsaturated zone (recharge from the unsaturated zone was an input to the model). Model results showed that P sorption processes (fast and slow) were the main loss pathways in groundwater. Fast sorption at the Cambridge site accounted for 86% of the groundwater losses. Fast sorption also dominated at the Muskoka site but slow sorption was more important (33% of loss). Some hydroxylapatite formed at the Cambridge site and a very small amount of stringite formed at the Muskoka site. The authors ran the model for 50 years into the future to investigate long-term effects. At the Cambridge site, the P front was expected to extend 50 m after 50 years. The P front didn't move at the Muskoka site due to low P concentrations in recharge entering the groundwater and high sorption within the aquifer. Over time, precipitate forms of P became a more important component of storage (the total P in the solid phase stayed the same). The authors also modeled what would happen if the source was cut off. This resulted in a pulse moving away from the drainfield. However, the concentration at the farthest boundary was about the same after 50 years. Robertson and Harman (1999) concluded that in decommissioned calcareous systems, down-gradient groundwater P concentrations are likely to persist for years.

Langergraber and Šimůnek (2005) developed a microbial growth model called CW2D to simulate N and P fate and transport in constructed wetlands (described in more detail in Section 3.2 on modeling N). It consisted of a multi-component reactive transport module that ran in the HYDRUS framework. The 13 components modeled in CW2D included inorganic P. Kinetic sorption of P and P release from mineralization of OM were simulated. A constant pH was assumed. Literature values for P sorption assuming a linear sorption isotherm (Equation 3.3.1-1) were given for three studies: K_d = 0.28, 19, and 185 cm^3/g. Values for nonlinear sorption of P were given for two studies: K_f = 38 and 58 $cm^{3\beta}/g^{\beta}$; $\beta = 0.83$ in both studies. No studies have been published on simulating P transport with CW2D, but Langergraber (2003) reported on a comparison between the model predictions and a tracer test using sodium chloride. It was necessary to assume that 5% of the water was immobile in a sand system in order to get a good fit.

Grant and Heaney (1997) developed a multi-component P transport module for the larger ecosystem model *ecosys*. The Gapon equation was used for competitive cation exchange reactions. Ten forms of soluble P were modeled. These and various forms of sorbed and precipitated P were associated with various P extractant (resin, $NaHCO_3^-$, NaOH, and HCl) fractions. The model simulations were compared to measurements of sorption isotherms and transport in a column of Ca-saturated cation exchange resin, a resin-sand column, and undisturbed soil columns. The model predicted the various forms of P accurately in most cases without calibration. Large changes in pH occurred and these changes affected P transformations. The authors noted that the multi-component transport approach has the advantage that it does not require soil-specific parameter data. However, the disadvantages are the requirement for more detailed information on soil chemical composition and the need for longer computer model simulation time to solve the chemical equilibria, as well as transport, system of simultaneous equations.

Another example of a multi-component transport model is the HP1 model (Jacques and Šimůnek, 2005) which is a combination of the PHREEQC geochemical model (Pakhurst and

Appelo, 1999) and the HYDRUS-1D model (Šimůnek et al., 2005). However, the researchers are not aware of any applications of this model to P fate and transport.

3.3.7 Other Phosphorus Models

Many comprehensive models of P fate and transport have been developed for agricultural conditions, in addition to the SWAT and REMM models. Brief descriptions of several other well known agricultural P models are included below:

- ANIMO is a comprehensive nutrient cycling model developed in The Netherlands to simulate leaching of P from the soil surface to groundwater and surface waters (Kroes and Roelsma, 1997). It simulates mineralization, sorption, plant uptake and leaching of P as well as N and oxygen concentrations and temperature in the soil profile.
- CENTURY (Grassland and Agroecosystem Dynamics Model) is designed to simulate long-term soil OM dynamics in the topsoil in response to changes in management and climate (Metherell et al.,1993). The model uses a monthly time step and simulates fluxes of C, N, and P.
- EPIC is a mass-balance model designed to simulate the effect of soil erosion on crop production (Williams, 1995). The P routines are very similar to those in SWAT (described in Section 3.3.3 above) because SWAT was developed from EPIC.
- RZWQM (Root Zone Water Quality Model) is a comprehensive agricultural model that simulates P fate and transport using numerical solutions (Ahuja et al., 2000).

3.3.8 Conclusions

Few P fate and transport models have been developed for OWTS and the processes that occur in the STU. Watershed-scale models may include estimates of P losses from OWTS to surface water bodies, but they emphasize the overland flow path and use relatively simple approaches to model P losses through runoff and leaching. In modeling the fate and transport of P within the STU, the P transformations of interest are mineralization, sorption, and the formation of precipitates. Since soils differ in clay mineralogy; Fe, Al, and Ca content; and pH, and these all affect sorption/precipitation, they differ substantially in their capacity to retain P. Studies have shown a two-stage retention process for soils with an initial fast, reversible reaction phase representing sorption, followed by a slow reversible or irreversible phase that may represent precipitation. Several questions are likely to be important in modeling P fate and transport in STUs:

- to what extent do sorption parameters vary among soils and how can these be related to soil properties;
- to what extent does preferential flow have an effect in allowing soluble P to bypass sorption sites and how can it be related to soil properties; and
- to what extent does sorption vs. precipitation occur?

The SWAT (Neitsch et al., 2002) and REMM (Altier et al., 2002) models employ similar algorithms to simulate P transformations and transport using a mass-balance approach. Different pools of organic P and sorbed P interact with dissolved P with fast and slow reactions for sorption/precipitation. REMM has equations for adjusting fast sorption for clay and pH and slow sorption for calcareous vs. noncalcareous soils. A number of P models have been developed based on use of the CDE to model one or more forms of soluble P. Several of these models assume two types of sites for P sorption: one site being a rapid or instantaneous sorption site and the other being a slow or kinetic site.

For a full treatment of sorption and precipitation, a multi-component reactive transport model that simulates fate and transport of the different forms of P and pH may be required. An example of this type of approach is the groundwater model developed by Spiteri et al. (2007) to describe sites in Canada where groundwater plumes from OWTS have been studied extensively. Phosphorus retention (sorption and precipitation) in the unsaturated zone at these sties removed 23% of effluent P at the calcareous site and 99% at the Fe-rich site (based on site measurements; the model did not include the unsaturated zone). Model results showed that P sorption processes (fast and slow) were the main loss pathways in groundwater. However, model simulations for 50 years into the future showed that precipitate-forms of P became a more important component of storage as P moved into mineral forms. It may be possible to adapt the multi-component transport model CW2D (Langergraber and Šimůnek, 2005) for P fate and transport in STUs. This comprehensive model could then be run for the different soil types to generate P loss data. Simple regression models could then be developed from this data that relate soil types to predicted P losses.

3.4 Modeling Microbial Transport in Onsite Wastewater Treatment Systems

As described in Section 2.5, porous media receiving STE act as natural filters that can remove microbial matter to variable extents. In general, pathogens carried with the percolating STE are most likely to be attenuated in subsurface (Lance and Gerba, 1984; U.S. EPA, 2001). In the study of the transport and fate of microbial matter, mathematical models can be useful tools for the quantitative assessment of microbial transport and fate in the subsurface. Modeling also permits elucidation of the individual transport parameters, such as sorption and inactivation, and their relative importance in controlling microbial concentrations in porous media. A quantitative understanding of microbial transport parameters in combination with transport models can then be applied by practitioners, researchers, government agencies and others to design and manage STE treatment systems.

The following discussion focuses on numerical models that describe the transport and fate of microorganism in the subsurface. The discussion, however, does not include numerical models like BIOPLUME, UTCHEM, or BIOMOC that simulate the transport and fate of compounds affected by the action of microorganisms, i.e., biodegradation and natural attenuation models. For a recent review of these models refer to Mulligan and Yong (2004) and Pavan and Worth (2008).

Baveye and Valocchi (1989) describe three conceptual frameworks that may be used for modeling bacterial growth and biologically reacting solute transport in saturated porous media. These include a biofilm model, a micro-colony model, and a macroscopic model. The biofilm model postulates that the solid particles constituting the transport domain are uniformly covered by a biofilm in which consumption of the substrate and electron acceptors takes place. Examples for the use of this model can be found in Bouwer and McCarty (1984) and Bouwer and Cobb (1987). The micro-colony model conceptualizes growth of microbial matter on discrete colonies of uniform constant dimensions (Molz et al., 1986). Based on this approach, microbial matter in the porous medium is accounted for by letting the number of micro-colonies per unit volume increase or decrease depending on substrate and electron acceptor utilization rate (Baveye and Valocchi, 1989). This micro-colony model was promulgated by Molz et al. (1986). The third type of model is characterized by the absence of any assumptions concerning the micro-scale configuration and distribution of the microorganisms in the pore space. Hence, in contrast to the biofilm and micro-colony model, this macroscopic model neglects pore scale processes, and the bacteria are assumed to respond to the macroscopic bulk fluid substrate concentration. One of the earlier attempts to utilize a macroscopic model was made by Borden and Bedient (1986). Because the geometric parameters describing the size of biofilms and micro-colonies are not readily available in the literature, macroscopic models, like those discussed in the following, are used most often (Pennington et al., 1999).

Modeling microbial matter has mostly focused on transport of viruses, which – relative to bacteria and protozoa – are the smallest particles. As such, viruses are least likely removed by mechanical filtration processes. In fact, as shown by Jang et al. (1983) and Gruesbeck and Collins (1982), straining of microbial matter is an important removal mechanism when the average particle size is greater than 5% of the grains that constitute the porous medium. For instance, the grain diameter of fine silt ranges from 2 to 6 μm, which is small enough to restrain the transport of most protozoa and some bacteria, but still large in comparison to virus matter. Because viruses are the most mobile particles and therefore of greatest concern in subsurface systems, the following review will focus mostly on modeling their transport.

Factors influencing the transport and fate of microbial matter in the subsurface include nature of the soil, microorganism type, attachment/detachment rates, filtration, temperature, microbial activity including the presence of other microorganisms, inactivation, moisture content, pH, OM content, and hydraulic conditions including advection-dispersion (Jin and Flury, 2002). Inactivation rate is considered the single most important parameter determining fate and transport of viruses in groundwater systems (Berger, 1994; John and Rose, 2005), and inactivation rate is often slower than the rate of die-off of other pathogens, such as infectious bacteria (Faulkner et al., 2003). However, there is not yet a consensus on which factor(s) have the greatest impact on eliminating microbial matter in general.

Modeling approaches to microbial transport vary greatly depending on the scale of the study and the specific interests of the investigators (Faulkner et al., 2002). Most mathematical models of microbial transport are based on the convection-dispersion equation (Hurst, 1991), which is expanded to include descriptive parameters such as attachment/detachment rates and bacteria die-off and unsaturated flow functions (e.g., van Genuchten, 1980). Some models approximate virus transport as a Fickian (i.e., diffusion-controlled) process, coupled with advective flow. Others apply colloid filtration theory to virus transport (Yao et al., 1971; Ryan and Elimelech 1996; Bhattacharjee et al., 2002; Tufenkji and Elimelech, 2004). Comprehensive reviews of the various modeling approaches, including tabulated transport and fate parameters, are presented by Schijven and Hassanizadeh (2000) and Jin and Flury (2002).

The removal of microbial matter in a specific soil can be characterized by calibrating transport models with laboratory or field data. However, the results depend on the conceptual basis of the model as well as the quality and availability of the input data (Corapcioglu and Haridas, 1984). Depending on the data input requirements of a particular transport model, some models are more useful than others for STE treatment system designers and managers. This is because most models have been developed for use in the saturated zone (see review in Azadpour-Keeley et al., 2003), whereas STE treatment relies processes in the unsaturated zone. In addition, these models often require input data that are unavailable at the field scale and are usually restricted to narrowly defined research systems. Those models that rely on a smaller number of readily available input parameters are typically limited to screening ("index models", Faulkner et al., 2003). Furthermore, there are a very limited number of field-test data sets available against which these transport models can be calibrated and validated (e.g., DeBorde et al., 1999; Saadoun et al., 2008).

3.4.1 Microbial Transport and Fate Models

In a review of existing mathematical models for virus transport simulation, Azadpour-Keeley et al. (2003) grouped models in two categories: readily available codes with user manuals, and codes for specific research purposes. The first group of virus transport models includes VIRALT, CANVAS, VIRTUS, VIRULO, and 3DFATMIC. Supporting information about these public-domain programs is available from the U.S. EPA's Center for Subsurface Modeling Support (CSMoS). The following discussion omits research models to focus on the well-documented, readily available models that are most likely to be used. The discussion of transport models is then extended to include a commercial model, HYDRUS, which has built-in microbial transport and fate modeling capabilities.

3.4.1.1 Public Domain Models

VIRALT The VIRALT code was developed for the U.S. EPA in 1994 and is a modular, semi-analytical and numerical code that simulates the single-source transport and fate of viruses in the saturated and unsaturated zones (Park et al., 1992). Processes considered in VIRALT are advection, dispersion, sorption, and virus inactivation (Azadpour-Keeley et al., 2003). VIRALT was developed to delineate ground-water pathlines and well capture zones, and to compute viral concentrations in extracted water. Contaminant sources of various shapes can be defined and the code computes viral concentrations in extracted water, providing both steady-state and transient transport. The one-dimensional transport equation is solved using the finite element method (FEM).

CANVAS The CANVAS code (Park et al., 1993) is a descendant of VIRALT and, in addition to the latter, can handle facilitated transport by colloidal matter as well as the simulation of multiple contaminant sources. CANVAS is a composite analytical-numerical FEM code that supports transient one-dimensional vertical flow and transport in the unsaturated zone and two-dimensional horizontal flow and transport in the saturated zone. Neither CANVAS nor VIRALT considers virus release (Bhattacharjee et al., 2002).

VIRTUS The "virus transport in unsaturated soils" (VIRTUS) code was developed by Yates and Ouyang (1992) as a predictive model of virus fate and transport in unsaturated soils that allows the virus inactivation rate to vary on the basis of changes in soil temperature. The model supports unsteady flow, transport in layered soils, different virus inactivation rates for adsorbed versus freely suspended virus particles, and the flow of heat through soils. It is assumed that viruses are introduced at the soil surface. VIRTUS is based on the law of mass conservation of a contaminant in porous media. It is a one-dimensional numerical finite difference code written in FORTRAN that simultaneously solves equations describing the flow of water, viruses, and heat through unsaturated soil under different climatic conditions (Yates and Ouyang, 1992). The equation governing the transport of viruses through the soil is given by:

$$\frac{\delta}{\delta t}\left(\rho_b C_S + \theta_m C_I\right) = \frac{\delta}{\delta z}\left(\theta_m D \frac{\delta C_I}{\delta z}\right) - V_I\left(\theta_m \frac{\delta C_I}{\delta z}\right) - \left(\theta_m \mu_I C_I + \rho_b \mu_S C_S\right) - \theta_m f\, C_I \tag{3.4.1.1-1}$$

where: ρ_b is the bulk density of the soil (M L^{-3}), C_s is the concentration of viruses adsorbed to the soil (in pfu per gram of solid), C_l is the concentration of viruses suspended in the liquid phase (in pfu per ml), D is the hydrodynamic dispersion coefficient (L^2 T^{-1}), V_l is the velocity of water in the liquid phase (L T^{-1}), θ_m is the volumetric soil water content (L^3 L^{-3}), μ_l is the inactivation rate of viruses in the liquid phase (T^{-1}), μ_s is the inactivation rate of adsorbed viruses (T^{-1}), f is the filtration coefficient (L^{-1}), t is time (T), and z is the downward distance (L). VIRTUS permits the user to input different virus inactivation rates for viruses that are adsorbed to the soil particles as compared with freely suspended viruses, if that information is known. The model was tested for its ability to predict virus movement measured in laboratory column studies. The model predictions, with the exception of one point, were within the 95% confidence limits of the measured virus concentrations (Yates and Ouyang, 1992). The VIRTUS program is in the public domain and can be obtained from the *International Ground Water Modeling Center* (IGWMC) at http://typhoon.mines.edu.

VIRULO The VIRULO code is a probabilistic screening model for predicting leaching of viruses in unsaturated soils. VIRULO employs Monte Carlo methods to generate ensemble simulations of virus attenuation due to physical, biological, and chemical factors. The model generates a probability of failure to achieve a user-chosen degree of attenuation (Faulkner et al., 2002). Conceptually, VIRULO is based on the work of Sim and Chrysikopoulos (2000), i.e.:

$$\frac{\delta[\theta_m C_l]}{\delta t} + \rho_b \frac{\delta C^*}{\delta t} + \frac{\delta[\theta_m C^o]}{\delta t} = D_z \theta_m \frac{\delta^2 C_l}{\delta z^2} - \frac{\delta[q C_l]}{\delta z} - \mu \theta_m C_l - \mu^* \rho C^* - \mu^o \theta_m C^o \tag{3.4.1.1-2}$$

where: $C_l = C(t, z)$ $(M L^{-3})$ is the concentration of viruses in the mobile solution phase, $C^*(t, z)$ $(M M^{-1})$ is the adsorbed virus concentration at the liquid-solid interface, $C^o(t, z)$ $(M L^{-3})$ is the adsorbed virus concentration at the liquid-air interface, q $(L T^{-1})$ is the specific discharge, θ_m $(L^3 L^{-3})$ is the moisture content, μ (T^{-1}) is the inactivation rate coefficient for the viruses in the bulk solution, μ^* (T^{-1}), the inactivation rate for the viruses that are sorbed at the liquid-solid interface, and μ^o (T^{-1}), the inactivation rate for the viruses sorbed at the liquid-air interface, ρ_b (ML^{-3}) is the soil bulk density, and $D_z = \alpha_z q/\theta_m + D_e$ $(L^2 T^{-1})$ is the hydrodynamic dispersion coefficient in the vertical direction, α_z (L) is the vertical dispersivity, $D_e = D/\tau$ $(L^2 T^{-1})$, where D $(L^2 T^{-1})$ is the virus diffusivity in water, and τ $(L L^{-1})$ is the tortuosity (>1). It is assumed that the simulated porous medium is homogenous in terms of hydraulic and virus properties as well as geochemistry. Further, the system is in steady state, only gravity drainage occurs, and preferential flow is not considered. Sorption and inactivation of viruses at the various interfaces are described by:

$$\rho \frac{\delta C^*}{\delta t} = k\,\theta_m \left(C_l - \frac{C^*}{K_d} \right) - \mu^* \rho_b C^* \tag{3.4.1.1-3}$$

where: $k = \kappa a_T$, $a_T = 3(1 - \theta_s)/r_p$ is the liquid-solid interfacial area in units of (L^2 L^{-3}). The symbol k (T^{-1}) is the microscopic mass transfer rate and κ (L T^{-1}) is called the mass transfer coefficient. In Equation 3.4.1.1-3, K_d (L^3 M^{-1}) is the equilibrium partitioning coefficient, r_p (L) is the average radius of soil particles, and θ_s (L^3 L^{-3}) is the saturated water content. The change in concentration of viable viruses at the air-water interface is calculated by:

$$\theta_m \frac{\delta C^o}{\delta t} = k^o \theta_m C - \lambda^o \theta_m C^o \tag{3.4.1.1-4}$$

where: k^o (T^{-1}) is the liquid-air interface mass transfer rate. The mass transfer rate for the liquid-air interface is described by:

$$k^o = \kappa^o\, a_T^o \tag{3.4.1.1-5}$$

where: κ^{o} (L^2L^{-3}) is the mass transfer coefficient and $a^{o}{}_{T}$ is the estimated area of the liquid-air interface as a function of the moisture content. The downward water flux in the unsaturated zone is modeled with the Buckingham-Darcy flow equation. Modeling parameters for Poliovirus as used in the development of the VIRULO model are provided in Table 3-2.

As pointed out by Azadpour-Keeley et al. (2003), virus transport modeling is inherently uncertain – often under-predicting virus transport – and therefore of limited use for determining regulatory compliance (Yates and Jury, 1995). Realizing this shortcoming, Faulkner et al. (2003) developed a screening model built upon governing equations in the VIRULO model, including equations for the advection, hydrodynamic dispersion, mass transfer, and inactivation of viruses in the unsaturated zone. Their model is not a fully dynamic virus transport model, but a screening model intended for predicting virus attenuation. The model attempts to minimize uncertainty in the prediction of attenuation by using Monte Carlo simulation in combination with parameters from the literature and independent simulation.

Table 3-2. Modeling Parameters for Poliovirus as Used to Develop the VIRULO Model (after Faulkner et al., 2002).

Parameter	n	Mean	Standard Deviation	Units	Source
$\log_{10}\lambda$	12	0.605	0.608	log(1/hr)	Breidenbach et al., 2009
$\log_{10}\lambda^*$	0	0.304	0.608	log(1/hr)	Chu et al., 2001
κ	1	1.34×10^{-3}	1.80×10^{-3}	m/hr	Yates and Ouyang 1992, assumed $\lambda^* \approx \lambda/2$
κo	1	9.27×10^{-3}	1.80×10^{-3}	m/hr	Yates and Ouyang 1992, assumed $\lambda^* \approx \lambda/2$
rv	0	1.375×10^{-8}	1.25×10^{-9}	m	Mazzone 1998, p. 114.
Kd (sand)	87	2.43×10^{-4}	5.66×10^{-4}	m^3/g	Breidenbach et al., 2009
Kd (silt loam)	23	3.77×10^{-43}	7.16×10^{-43}	m^3/g	Breidenbach et al., 2009
Kd (clay)	39	7.20×10^{-4}	9.74×10^{-4}	m^3/g	Breidenbach et al., 2009

Note: λ is the suspended phase inactivation rate, λ* is the inactivation rate for the viruses that are sorbed at the liquid-solid interface, κ is the liquid-solid mass transfer coefficient, κ^{o} is the liquid-air mass transfer coefficient, rv is the virus radius, and Kd is the equilibrium partitioning coefficient.

3DFATMIC The 3DFATMIC is a three-dimensional subsurface transport and fate model developed to simulate transient and/or steady-state density-dependent flow field and transient and/or steady-state distribution of a substrate, a nutrient, and an aerobic or anaerobic electron acceptor in a three-dimensional domain of subsurface media. The code is based on the generalized Richards equation and Darcy's law which are simulated with the Galerkin FEM (Yeh et al., 1997). Chemical and bacteria transport are simulated based on the principle of conservation of mass and Monod kinetics. According to Yeh et al. (1997), this model can completely eliminate peak clipping, spurious oscillation, and numerical diffusion. Pennington et al. (1999) used 3DFATMIC to simulate the interaction of explosives, their by-products, and biomass. However, because of limited input data availability and problems with the numerical code, the model runs were not usable. The 3DFATMIC model is distributed by the Scientific Software Group, Sandy UT, but the UNIX source code can be obtained from CSMoS free of charge.

3.4.1.2 Commercial Model

HYDRUS The HYDRUS software package is a commercially available, Microsoft Windows-based modeling environment for analysis of water flow, solute, and heat transport in variably saturated porous media. The software package consists of the computational computer program, and the interactive graphics-based user interface. There are HYDRUS versions available with one-, two- and three-dimensional transport modeling capabilities. The HYDRUS program numerically solves the Richards equation for variably saturated water flow and advection-dispersion equations for both heat and solute transport. Attachment/detachment theory, including filtration theory, is included to permit simulation of the transport of viruses, colloids, and/or bacteria (Šimůnek et al., 2007). Virus, colloid, and bacteria transport and fate is simulated according to Equation 3.4.1.2-1 (Šimůnek et al., 2006):

$$\frac{\delta \theta C_l}{\delta t} + \rho_b \frac{\delta C_{se}}{\delta t} + \rho_b \frac{\delta C_{s1}}{\delta t} + \rho_b \frac{\delta C_{s2}}{\delta t} = \frac{\delta}{\delta x_i}\left(\theta_m D_{ij}^w \frac{\delta C_l}{\delta x_j} \right) - \frac{\delta q_i C_l}{\delta x_i} - \mu_w \theta C - \mu_s \rho \left(C_{se} + C_{s1} + C_{s2} \right) \quad (3.4.1.2\text{-}1)$$

where: C_l is the (colloid, virus, bacteria) concentration in the aqueous phase ($N_c\, L^{-3}$), C_s is the solid phase (colloid, virus, bacteria) concentration ($N_c\, M^{-1}$], subscripts e, 1, and 2 represent equilibrium and two kinetic sorption sites, respectively, N_c is a number of colloids (particles), D_{ij}^w is the dispersion coefficient tensor for the liquid phase ($L^2\, T^{-1}$), x_i are the spatial coordinates (i=1,2,3) (L), q_i are the components of the Darcian fluid flux density ($L\, T^{-1}$), and μ_w and μ_s represent inactivation and degradation processes in the liquid and solid phases (T^{-1}), respectively.

Microbial mass transfer between the aqueous and solid kinetic phases is described by:

$$\rho_b \frac{\delta C_s}{\delta t} = \theta k_a \psi C_l - k_d \rho_b C_s \quad (3.4.1.2\text{-}2)$$

where: ψ is a dimensionless colloid retention function (-), k_a is the first-order deposition (attachment) coefficient (T^{-1}] and k_d is the first-order entrainment (detachment) coefficient (T^{-1}). According to Šimůnek et al. (2006), the attachment and detachment coefficients are strongly dependent upon water content, with attachment significantly increasing as the water content decreases. The attachment coefficient is calculated using filtration theory (Logan et al., 1995), which is a quasi-empirical formulation in terms of the median grain diameter of the porous medium ("collector"), the pore-water velocity, and collector and collision ("sticking") efficiencies accounting for colloid removal due to diffusion, interception and gravitational sedimentation (Rajagopalan and Tien, 1976; Logan et al., 1995).

A modified HYDRUS-1D code (Šimůnek et al., 1998) was used by Bradford et al., (2006) to simulate the transport and deposition of *Giardia* cysts and manure suspension in column experiments. The authors adapted HYDRUS-1D to account for colloid attachment, detachment, straining, and size exclusion and coupled the model with a least square optimization routine. Their results show that *Giardia* transport in their columns filled with Ottawa sands (d_{50} = 150 to 710 μm) was primarily controlled by straining.

Pang et al. (2008) used the HYDRUS-1D (Šimůnek et al., 2005) and a two-region mobile–immobile water model (MIM) to evaluate the transport of fecal coliforms, *Salmonella* bacteriophage, and bromide tracer in ten soils, ranging from clayey gley soils to gravel. Modeling results indicated exclusion of microbes from smaller pores. In addition, the detachment rate was only 1% of the attachment rate, indicating irreversible attachment of microbes. Soil structure (macro-porosity) appeared to play the most important role in the transport of microbes and bromide tracer, while soil lithology had the greatest influence on attenuation and mass exchange. The MIM was utilized to calculate removal rates. MIM assumes that water flow and contaminant transport is limited to the mobile water region and that water in the immobile water region is stagnant, with a first-order diffusive exchange process between the two regions. The microbial reduction varied greatly, but was 2 to 3 orders of magnitude per meter for most soils.

Richardson and Janna (2006) analyzed the hydraulic performance of two sizes of subsurface wastewater drainage systems (trenches) with various dosage levels using HYDRUS-2D. These systems were modeled to determine the performance when first installed, and again when they reached maturity. It was found that as the dosing rate increased the ponding height of the trenches increased for all systems, although not linearly. As all systems matured, the flow rate out of the system decreased and the ponding height increased.

3.4.2 Outlook

The review of existing public-domain and commercial models demonstrate that simulation of microbial characteristics in OWTS is still largely uncharted territory. According to Faulkner et al. (2003), progress has been made to measure parameters to integrate microscopic filtration theory into models describing virus transport in column studies. However, their use in screening models to predict leaching is often impractical. This is due to a lack of knowledge of the aggregated geochemical and physico-chemical properties of soils in most sites of interest to planners. Furthermore, the geochemistry of soil mineral surfaces is known to play a significant role in the adsorption behavior of percolating viruses (Faulkner et al., 2003). Yet again, the effects of geochemical heterogeneity on microbial transport are not well understood at this time. As demonstrated by Bradford et al. (2006), the presence of other contaminants, such as manure or surface active compounds (Chattopadhyay et al., 2002) further complicates the simulation of microbial transport. In addition, shock loads of high microbe concentrations are likely to occur in STE treatment units, possibly altering the soil's ability to remove microbial matter. Finally, heterogeneous soil structures, i.e., presence of aggregation in soils, are still difficult to model accurately. Nonetheless, interactions between microbial matter and inert soil constituents apparently influence inactivation rates. Although most of these processes are currently not included in commonly available microbe transport and fate codes, research is underway to better understand the these processes and include them in future models.

3.5 Models Used to Predict Fate and Transport Processes of OWCs

While OWC studies have focused on the appearance and concentration of trace organic compounds in wastewater, treated effluent, and environmental systems, several studies have been conducted to model fate and transport. In considering a model, it is important to accurately represent the relevant transport and elimination processes in a specific system. OWCs form a broad class of compounds. Thus, it is likely that no single model or modeling approach will be appropriate for all OWCs, or even for the five target OWCs (triclosan, 4-nonylphenol, 1,4-dichlorobenzene, 17β-estradiol, and diclofenac) chosen for this research. This chapter reviews the models that have been used to simulate transport and fate of OWCs. The review includes fate and transport modeling software, fate and transport equations, and equations to model specific processes and relationships.

In a study by Keller (2005), the following models were investigated for in-stream removal and in WWTP and wastewater conveyance system elimination of "down-the-drain" chemicals such as detergents and triclosan. The study investigates one fugacity model, one hydrodynamic model (Mike 11), one steady state model (QUAL-2E), and two stochastic models (TOMCAT and GREAT-ER).

A fugacity model relates chemical potential in the form of adjusted pressure to the tendency of a substance to prefer one phase over another. Application of fugacity models for this work is described in Mackay et al. (1992). Mike 11 is a system for the 1-dimensional, dynamic modeling of rivers, channels and irrigation systems, including rainfall runoff, advection-dispersion, morphology, and water quality (DHI, 2003). It is commonly used to predict in-stream concentrations and as a water quality management tool (Thompson et al., 2004). QUAL2E is a comprehensive and versatile stream water quality model. It can simulate up to 15 water quality constituents in any combination desired by the user. Constituents that can be simulated are DO, BOD, temperature, algae, N, P, coliform bacteria, and up to three conservative and one nonconservative constituents (Bowden and Brown, 1984). TOMCAT is based on the hybrid Monte-Carlo deterministic method and simulates flow distributions and concentrations of key determinants including BOD, ammonia, DO, and temperature (Crockett et al., 1989). GREAT-ER is a model for environmental risk assessment and management of chemicals in river basins. The program deals with geo-referenced datasets instead of averaged or generic values. It has been applied to study the aquatic fate of cleaning agent and detergent chemicals in the Ruhr River catchment (in Germany) and the UK (Boeije et al., 2000; Fox et al., 2000; Matthies et al., 2001; Sabaliunas et al., 2003).

The authors concluded that all of these models do a good job of predicting in-stream removal. Fugacity models are more appropriate for removal in sewage in WWTPs and through sedimentation in streams. Mike II and QUAL2E are more data-intensive than TOMCAT and GREAT-ER. The authors concluded that when the large amount of detailed input data required for Mike II and QUAL2E is not available, GREAT-ER is the most appropriate tool for modeling at catchment scale because, unlike TOMCAT, it is able to model the sewage treatment component.

HYDRUS 2D/3D is a computational finite element model that can be used to simulate the two and three dimensional movement of water, heat, and multiple solutes in variably saturated media. The model includes a parameter optimization algorithm for inverse estimation of a variety of soil hydraulic and/or solute transport parameters (Šimůnek et al., 1991). The program has been used to predict breakthrough curves for homogeneously packed and layered columns for hormones (including 17β-estradiol, estrone, and testosterone) (Das et al., 2004).

The CXTFIT code by van Genuchten and Parker (1984) modified by Toride et al. (1995) was used to perform simulations by Scheytt et al. (2006). This code accounts for factors that contribute to changes in the hydrodynamic properties of sediment under unsaturated conditions. One of these factors is the existence of regions of immobile water versus regions of mobile water. A wider spectrum of pore water velocities is considered when media is unsaturated and tortuosity of the solute flow path increases as media saturation decreases. Scheytt et al. (2006) applied this model to the pharmaceuticals carbamazepine, diclofenac, ibuprofen, and propyphenazone.

McAvoy et al. (2002) provides a general model for nonpolar, more volatile, OWCs, such as 1,4-dichlorobenzene, through septic tanks, the unsaturated zone, and the saturated zone. The model assumes sorption and biodegradation are the main factors affecting transport. The model uses simple mixing equations and/or solutions to the advection-dispersion equation (with linear, equilibrium reactions) that can be implemented in a spreadsheet. This model, therefore, is most appropriately used as a screening model. The equations utilized in the model are listed below:

The Septic Tank Module includes Equations 3.5.1-1 through 3.5.1-3:

$$STE = \frac{STI}{\left[\left(k_{b1} f_d \tau \right) + \left(\frac{v_s f_p \tau}{H} \right) + 1 \right]} \qquad (3.5.1\text{-}1)$$

$$f_d = \frac{1}{k_{d1}(M+1)} \quad (3.5.1\text{-}2)$$

$$f_p = 1 - f_d \quad (3.5.1\text{-}3)$$

where: STE (mg/L) = total effluent concentration, STI (mg/L) = total influent concentration, k_{b1} (1/day) = 1st order anaerobic biodegradation rate, v_s (m/d) = suspended solids settling velocity, f_d = fraction of compound in dissolved phase, f_p = fraction of compound sorbed, H (m) = effective depth of settling tank, τ (d) = hydraulic retention time, K_{dl} = sorbed-aqueous phase distribution coefficient (L/kg), and M = suspended solids concentration of wastewater (kg/L).

The Unsaturated Zone Module includes Equations 3.5.1-4 through 3.5.1-7:

$$C_z = C_0^{-\lambda z} \quad (3.5.1\text{-}4)$$

$$\lambda = R_2 \frac{k_{b2}}{v} \quad (3.5.1\text{-}5)$$

$$R_2 = 1 + K_{d2}\left(\frac{\rho_{b2}}{\eta S_w}\right) \quad (3.5.1\text{-}6)$$

$$v = \frac{Q}{\eta S_w A} \quad (3.5.1\text{-}7)$$

where: $C_{(z)}$ (mg/L) = dissolved concentration at depth z, C_0 (mg/L) = dissolved concentration at zero depth, R_2 = retardation factor for unsaturated zone, z (m) = depth below drainage field, η = effective porosity of the unsaturated soil, S_w = relative saturation of the unsaturated soil, ρ_{b2} (kg/L) = bulk density of unsaturated soil, K_{d2} (L/kg) = soil-water distribution coefficient for the unsaturated soil, k_{b2} (1/day) = 1st order biodegradation rate, v (m/d) = water velocity, A (m^2) = effective drainage field area, and Q (m^3/d) = drainage field flow.

The Saturated Zone Module includes Equations 3.5.1-8 through 3.5.1-11:

$$C_{x,t} = \frac{C_0}{2}\exp\frac{(v-u)x}{2D}\, erfc\frac{R_3 x - ut}{2(R_3 D t)^{0.5}} + \exp\frac{(v+u)x}{2D}\, erfc\frac{R_3 x + ut}{2(R_3 D t)^{0.5}} \quad (3.5.1\text{-}8)$$

$$v = K\frac{i}{\eta} \quad (3.5.1\text{-}9)$$

$$R_3 = 1 + K_{d3}\frac{p_{b3}}{\eta} \quad (3.5.1\text{-}10)$$

$$u = v\left[1 + (4k_{b3}\frac{D}{v^2}\right]^{0.5} \quad (3.5.1\text{-}11)$$

where: $C_{(x,t)}$ (mg/L) = dissolved solute concentration at distance x and time t, C_0 (mg/L) = dissolved solute concentration at water table, v (m/d) = groundwater velocity, u (m/d) = effective velocity of the solute, x (m) = longitudinal distance in soil, R_3 = retardation factor for the soil, t (d) = time, D (m^2/d) = hydrodynamic dispersion coefficient for saturated soil, K (m/d) =

hydraulic conductivity, i (m/m) = hydraulic gradient, η = effective porosity, K_{d3} (L/kg) = distribution coefficient for the soil, ρ_{b3} (kg/L) = bulk density of the soil, and k_{b3} (1/day) = 1st order biodegradation rate in saturated soil.

Das et al. (2004) used the advection-dispersion equation with non equilibrium, dual-domain sorption and first-order transformation rates to model fate and transport of testosterone and 17β-estradiol in sediment. Over the transport distance, water content and other soil properties are homogeneous. Equations 3.5.1-12 through 3.5.1-16 are the governing transport equations for this model (saturated or unsaturated zone):

$$\beta R\left(\frac{\delta C_1}{\delta T}\right)=\frac{1}{P}\left(\frac{\delta^2 C_1}{\delta Z^2}\right)-\frac{\delta C_1}{Z}-\omega\left(C_1-C_2\right)+\gamma_1 C_1 \tag{3.5.1-12}$$

$$\beta R\frac{\delta C_2}{\delta T}=\omega\left(C_1-C_2\right)+\gamma_2 C_2 \tag{3.5.1-13}$$

$$\beta=\frac{\theta+f\rho K_d}{\theta+\rho K_d} \tag{3.5.1-14}$$

$$R=1+\frac{\rho K_d}{\theta} \tag{3.5.1-15}$$

$$T=\frac{vt}{L} \tag{3.5.1-16}$$

where: β = dimensionless parameter related to the fraction of equilibrium sorption sites, R = retardation factor, T = dimensionless time, $C_1 = C/C_0$, dimensionless aqueous concentration, C_2 = dimensionless sorbed-phase concentration, γ_1 = dimensionless rate coefficient for degradation in liquid phase, γ_2 = dimensionless rate coefficient for degradation in solid phase, ω = Damkohler number, θ = water content, f = fraction of equilibrium sorption sites, ρ (g/mL) = soil bulk density, K_d = linear sorption coefficient, v (cm/s) = pore water velocity, t (s) = time, and L (cm) = soil column length or transport distance.

Where sorption is considered in the reviewed models, all assume that the soil-water partition coefficient is related to the organic carbon content of the soil material. The organic carbon partition coefficient is calculated based on the octanol-water partition coefficient using various log relationships in many studies examining OWCs. Equations 3.5.1-17, 3.5.1-18, and 3.5.1-19 are used by Wilson et al., (1981) to evaluate halogenated hydrocarbons and substituted benzenes. Equation 3.5.1-20 is used by Zwiener and Frimmel (2003) to evaluate clofibric acid, ibuprofen, and diclofenac. Equation 3.5.1-21 is used by Schwarzenbach and Westall (1981) to evaluate volatile organics such as 1,4-dichlorobenzene.

$$\log G = 4.04 - 0.557\log WS_1 \tag{3.5.1-17}$$

$$\log K_{oc} = -0.54\log WS_2 + 0.44 \tag{3.5.1-18}$$

$$\log K_{oc} = 3.64 - 0.55\log WS_3 \tag{3.5.1-19}$$

where: WS_1 (µmol/L) = water solubility, WS_2 (mol fraction) = water solubility, WS_3 (mg/L) = water solubility, and G = partition coefficient based on OM (K_{oc} = G/1.724).

$$\log K_{oc} = 0.72\log K_{ow} + 0.49 \tag{3.5.1-20}$$

$$\log K_p^z = 0.72 \log K_{ow}^z + \log f_{oc} + 0.49 \quad (3.5.1\text{-}21)$$

where; K_p^Z = partition coefficient for nonpolar compound Z, K_{ow}^Z = octanol-water partition coefficient for Z, and f_{oc} = fraction of organic carbon.

Linear sorption isotherms are commonly used but many studies found the non-linear Freundlich isotherm better fit data for certain OWCs (Poole and Poole, 1999; Ullman, 2007). Ullman (2007), studying environmental transport of the hormones 17β-estradiol and estrone, found that the Langmuir isotherm fit best for loamy fine sand. All reviewed studies modeled biodegradation (aerobic and anaerobic) as a first-order reaction.

Sorption and biodegradation are two major processes that affect OWC fate and transport but there are a number of other processes that should be considered. Transformation by photolysis, hydrolysis, and volatilization are dominant processes for some OWCs. Buser et al. (1998) and Tixier et al. (2003) modeled the phototransformation of diclofenac. Sorption of neutral organic compounds to soil from the gas phase was modeled by Poole and Poole (1999). This gas phase sorption model is:

$$\log SP = c + 1 \log L^{16} + rR_2 + s\pi_2^H + a\sum \alpha_2^H + b\sum \beta_2^H \quad (3.5.1\text{-}22)$$

$$\log K_{aw} = \log K_s + \log L^w \quad (3.5.1\text{-}23)$$

where: SP = K_{aw} (air water partitioning coefficient), K_{aw} (cm) = ratio of the surface concentration (mol/cm^2) to the molar gas phase concentration (mol/cm), K_s (cm) = ratio of the surface concentration (mol/cm^2) to the bulk aqueous phase concentration (mol/cm^3), L^w = the air-water distribution constant for organic vapors, L^{16} = distribution constant for the solute between gas and n-hexadecane, R_2 (cm^3/10) = excess molar refraction, π_2^H = the solute's dipolarity/polarizability, $\Sigma\alpha_2^H$ = the solute's hydrogen bond acidity, $\Sigma\beta_2^H$ = the solute's hydrogen bond basicity, r = constant that determines the difference in capacity of the wet soil to interact with solute n electrons, s = constant to the difference in the capacity of wet soil and water or air to take part in dipole-dipole and dipole-induced dipole interactions, a = constant to the difference in hydrogen-bond basicity of wet soil and water or air, b = constant to the difference in hydrogen bond acidity of the wet soil and water or air, and l = constant to measure ease of forming cavity for the solute in wet soil.

The effects of environmental conditions are considered in some models. The influence of pH on in-stream removal of triclosan was investigated by Sabaliunas et al. (2003). The pH of the stream water increased only from 6.94 to 7.18. Because triclosan is a sorptive and ionizable compound, slight changes in pH and suspended solids concentration may cause repartitioning between the sorbed and aqueous phase. Equations 3.5.1-24 through 3.5.1-26 are included in this model for in-stream removal:

$$C_{sol-union} = \frac{C_T}{1 + 10^{pH-pKa} + K_{oc} f_{oc} M} \quad (3.5.1\text{-}24)$$

$$C_{sol-ion} = C_{sol-union} 10^{pH-pKa} \quad (3.5.1\text{-}25)$$

$$C_{sor-union} = K_{oc} f_{oc} M C_{sol-union} \quad (3.5.1\text{-}26)$$

where: $C_{sol\text{-}union}$ (mg/L) = concentration of dissolved unionized triclosan, $C_{sol\text{-}ion}$ (mg/L) = concentration of dissolved ionized triclosan, $C_{sor\text{-}union}$ (mg/L) = concentration of sorbed compound, C_T (mg/L) = total concentration, K_{oc} (L/kg) = water-organic carbon partition coefficient, f_{oc} = fraction of organic carbon in suspended solids, and M (kg/L) = suspended solids concentration.

The effect of salinity on biodegradation of nonylphenol was modeled by Jonkers et al., (2005). The relationship between degradation rates and salinity in estuaries was empirically derived according to:

$$\lambda_{bd} = \lambda_{op}\left[3 - 3.07\left(\frac{salinity}{35}\right)^{0.5}\right] \qquad (3.5.1\text{-}27)$$

where: λ_{bd} (day^{-1}) = degradation rate, and λ_{op} (day^{-1}) = optimum degradation rate.

This review provides examples of different modeling software, transport equations, and process-specific equations and relationships. The models included consider homogenous media and linear, equilibrium reactions are typically assumed for sorption. Kinetic sorption has not been considered. Biodegradation is modeled using first-order kinetics in all models. Geochemistry related reactions, such as oxidation-reduction, are generally not considered.

Evaluating the five contaminants selected for this study—triclosan, 4-nonylphenol, 1,4-dichlorobenzene, 17β-estradiol, and diclofenac—will require models to predict fate and transport under various soil conditions. Processes to consider for modeling include sorption, abiotic degradation (hydrolysis, photolysis), and biodegradation (anaerobic, aerobic).

CHAPTER 4.0

SUMMARY AND CONCLUSIONS

4.1 Summary

The literature review described in this report is part of a larger research project to assess STU performance with respect to treatment of important wastewater constituents. The overall goal of the project is to provide a toolkit and tool-use protocol that is easy to implement and available to a wide range of users to assess STU performance. All tools developed will be based on rigorous experimental data and quantitative models verified with field data from operating systems. In some cases, more sophisticated tools (e.g., complex mathematical models) may be warranted depending on the complexity of the problem and the relative risk associated with a poorly designed STU.

Over 200 data points from 84 experiments where pulled from the literature that describe N treatment and removal within STU. Only 11 sources describing 30 experiments were found that report P concentrations with depth in STUs. The majority of available peer-reviewed studies for microbial fate and transport have been conducted at laboratory or pilot scales while field-scale evaluations are limited. Finally, little research has been published on treatment and removal processes for OWCs in STUs; most studies of OWC focus on the occurrence in surface water and groundwater of selected compounds from municipal wastewater treatment plants. Data from the literature were compiled and analyzed to determine key conditions that affect STU performance. The literature review also focused on best practices for using models and other tools to predict STU performance. The information gained during this literature review will direct the design of field experiments to be conducted at CSM, URI, and UGA as needed to fill any data gaps and/or develop the design tools to assess STU performance.

4.2 Conclusions

4.2.1 Soil Treatment Unit Performance

Based on the findings reported in the literature, several conclusions can be drawn related to STU performance:

- The factors that lead to denitrification, or removal of nitrates from the soil, are well understood and include: the presence of a carbon source, oxygen diffusion, soil moisture, and temperature. However, these factors are not generally measured or reported in the literature (especially for field experiments).
- Nitrogen in soil pore water was often reported at concentrations higher than the concentration in the applied wastewater. Because of the relatively large numbers of C/C_o values that are greater than 1.0, it is not likely that this is due to analytical error.
- Given the variability of data collected at field sites, simple binary relationships (e.g., C/C_o versus depth for various soil types) for N attenuation based on literature values are not likely to produce statistically justifiable predictions. Preliminary analysis suggests that multivariate statistical methods are more likely to produce reliable predictive tools. Thus, multivariate

statistical analysis will be conducted between N treatment and all the factors that might influence removal (e.g., depth, soil texture, HLR, water content, and other soil conditions).

- In field studies, HLR appears to be more important than soil texture and soil depth with respect to N treatment or depth for field soils (based on literature data from 30-60 depth interval), likely due to the inherent variability of soil properties at the field scale. In contrast, soil type was the most important variable for predicting N removal in laboratory experiments.
- STE-derived P is involved in several chemical processes in the STU, including degradation of polyphosphates to orthophosphates, adsorption, and chemical precipitation. Most of the P-soil interaction data found in the literature appear to be in the form of batch tests. Field conditions with high HLRs, nonuniform distribution of STE in the STU, and the lack of a biozone within the trenches are poorly studied or reported in the literature.
- Little information exists in the literature to determine the P retention in the STU based on soil type, mostly because P attenuation processes are not fully understood and difficult to model.
- The majority of available peer-reviewed studies have been conducted at laboratory or pilot scales. Field-scale evaluations of pathogen removal are scant, and they focus on a particular type of organism, with little or no consideration of other functions performed by the STU.
- There are clear differences in factors controlling the fate of different types of pathogenic organisms found in STE which present a challenge for optimization of removal of these organisms in conventional soil-based systems.
    - Soil texture and treatment depth do not appear to be useful for developing statistical tools to predict viral fate. The interplay of virus isoelectric point, pH and clay mineralogy appears to be an important factor in determining virus removal, as are the level of dissolved OM in STE and the presence of unsaturated conditions below the infiltrative surface. Lower HLRs should improve virus removal, primarily because soil-water contents would be lower.
    - Removal of bacterial pathogens takes place primarily by mechanical filtration, a process that is governed by soil texture, treatment depth, and the presence of unsaturated conditions below the infiltrative surface. Hydraulic loading rate does not appear to have a consistent effect on bacterial removal.
    - Mechanical filtration appears to be the main process for removal of protozoan cysts and oocysts from STE.
- Studies in peer-reviewed literature have focused on the occurrence and concentrations of selected compounds in streams, lakes, and groundwater impacted by wastewater. Most of the existing work focuses on scenarios where the OWC source to the environment was wastewater treatment plant effluents. Future research is required for a better understanding of fate and transport processes of these compounds as they are integrated into the environment with respect to removal in OWTS STUs.
- The most relevant removal processes for OWCs in STUs are sorption, biodegradation, and volatilization. Little research has been conducted in these areas.
- Some commonalities are apparent with respect to removal of various constituents in OWTS:
    - The presence of organics in soil water improves N removal because it is necessary for denitrification, but degrades treatment of virus in soil. Thus, OWTS operations that

remove organic material from effluent before it is applied to the soil may inhibit soil denitrification, but enhance virus removal. In addition, while it was not discussed in the literature, it is possible that colloidal organic matter can facilitate transport of hydrophobic OWCs that may sorb to organic colloids.

  - Low water saturation (and high air saturation) in soils enhances nitrification, removal of virus (via enhanced partitioning to the water-air interface) and removal of OWCs that degrade aerobically. However, low water saturations will prevent removal of nitrogen via denitrification. On the other hand, some OWCs (that are not part of this study) degrade anaerobically, so higher water contents would improve removal of these contaminants. Oxygen promotes irreversible chemical precipitation of P, the primary removal mechanism for this constituent. Thus lower water contents would improve removal of P. In general, higher HLR will result in higher water contents.
  - The theoretical and statistical evaluations described in this report generally suggest that lower HLR should improve removal of N, virus, and phosphorus.

- In many cases, statistical analysis of literature data suggests that factors such as soil depth or soil type are not useful for developing predictive statistical tools for removal of most wastewater constituents. However, we know from fundamental theory that these factors are important for removal of most wastewater pollutants. Thus, while these factors may not be useful for development of statistical tools, they still must be considered when developing modeling tools that can be applied across many different sites where data are not available.

4.2.2 Soil Treatment Unit Modeling Tools

Based on the findings reported in the literature, several conclusions can be drawn related to available modeling tools applicable to OWTS:

- Many N models have been developed for agricultural applications but very few have been developed for OWTS and the processes that occur in the STU.
- The critical question in modeling N in OWTS is under what conditions and to what extent does denitrification occur. One HYDRUS N model adapted for OWTS was most sensitive to the denitrification rate, and literature values for this input parameter vary considerably.
- Several studies indicate that differences in soil texture, structure or drainage class are likely to affect denitrification, largely through their effect on soil water and oxygen availability.
- It may be possible to develop simple N models for STUs that will predict the effect of different soil types (texture, structure, and drainage class) on N removal by adapting the CW2D model to OWTS. CW2D is a HYDRUS-based model of a sand filter that incorporates most of the features needed in a comprehensive microbial growth model, including a variable rate of denitrification due to changes in DO concentrations. The model DrainMod-N-II also appears promising. However the latter model is not well supported, and is one-dimensional, and thus cannot simulate many scenarios related to OWTS design.
- Few P fate and transport models have been developed for OWTS and the processes that occur in the STU. Watershed-scale models may include estimates of P losses from OWTS to surface waterbodies, but they emphasize the overland flow path and use relatively simple approaches to model P losses through runoff and leaching.
- Very few models and laboratory experiments associated with P consider both soil sorption and precipitation, even though these processes are known to be very important to P transport.

Thus, understanding of P fate and transport at the field scale is very limited. Generally, P pollution from OWTS is not a concern for most geographic locations.

- As with N modeling, it may be possible to adapt the multi-component transport model CW2D for P fate and transport in STUs.
- The review of existing public-domain and commercial models demonstrates that simulation of microbial characteristics in OWTS is still largely uncharted territory. Integration of microscopic filtration theory into models describing virus transport in column studies in screening models to predict leaching is often impractical. The effects of geochemical and soil structure heterogeneity on microbial transport, and high microbe concentrations that are likely to occur sporadically in STE treatment units, are not well understood at this time. However, research is underway to better understand these processes and include them in future models.
- Several studies have been conducted to model OWC fate and transport. However, OWCs form a broad class of compounds, making it unlikely that a single model or modeling approach will be appropriate for all OWCs. Processes to consider for modeling include sorption, abiotic degradation (hydrolysis, photolysis), and biodegradation (anaerobic, aerobic).
- No single model exists that is appropriate for modeling all wastewater constituents considered in this study.

APPENDIX A

OWC SUMMARY AND FACT SHEETS

Table 8	Removal Processes (dominant process for contaminant in **bold**)						Bioconcentration
Contaminant	Aerobic Biodegradation	Anaerobic Biodegradation	Adsorption	Photolysis	Hydrolysis	Volatization	Degree of BCF
Triclosan (TCS)	**Highly biodegradable with half lives ranging from 15 to 35 days**	Little to no anaerobic biodegradation	Moderate to high adsorption to sludge. Sorption in soils increases with organic carbon content.	Important for soil water pH above 8 only. Relevant removal process for lake water.	Not noted for TCS	Due to low Henry's constant, not expected	moderate to low
Methyl-triclosan (MTCS)	Biotic daughter product of TCS. Does not biodegrade as readily as TCS.	Little to no anaerobic biodegradation	**Moderate to high adsorption to sludge. Sorption in soils increases with organic carbon content.**	Not noted for MTCS	Not noted for MTCS	Due to low Henry's constant, not expected	very high
4-nonylphenol (NP)	**Fairly biodegradable with 90% removal shown in 40 to 320 days in soil. 1/2 life reported at 2.5 days, 80% removal in 35 days**	Low anaerobic biodegradation. Closed flask 1/2 life = 16 days	High adsorption to sludge. Sorption in soils increases with organic carbon content	Not noted for NP	Not noted for NP	Due to low Henry's constant, not expected	low
1,4-dichlorobenzene	Moderate in soil water	Not expected to occur	Moderate to low adsorption.	Expected to be low in soil water	Expected to be low in soil water	**Expected to be high in soil water**	moderate to high
17β-estradiol	Estimated degradation rate coefficient: 0.003 to 0.075/hr	little anaerobic biodegradation	**High adsorption. Increases in the presence of bioactivity. Dissolved organic carbon has no significant effect.**	Not noted	not noted	Due to low Henry's constant, not expected	very high
Diclofenac	Not significant	Not expected to occur	Low adsorption	Extremely effective	May be significant	Due to low Henry's constant, not expected	very high

A.1 Triclosan Fact Sheet

Molecule Image:

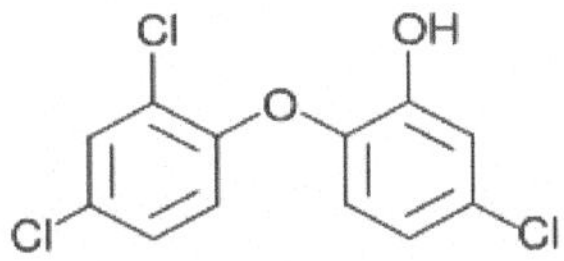

Chemical Formula:
C12H7Cl3O2

Uses:
Triclosan is found in soaps (0.15-0.30%), deodorants, toothpastes, shaving creams, mouth washes, and cleaning supplies and is infused in an increasing number of consumer products, such as kitchen utensils, toys, bedding, socks, and trash bags, sometimes as the proprietary Microban treatment.

Occurrence:

Water Source	Median Concentration (µg/L)	Maximum (µg/L)	Sources
Wastewater	1.3	4.4	Conn, 2006
Septic Tank Effluent	.83 res 4.2 nonres	9.3 res 82 nonres	Conn, 2006
River water	0.14	2.3	Kolpin, 2001
Lake water	0.01	0.012	Lindstroem, 2002

Chemical Properties:

Parameter	Value Range	Sources
log Koc	4.265 - 4.7	Ying, 2007
Solubility (mg/L)	5.0 - 10.0	Sietz, 2005
Kh (atm-m3/mole)	1.52E-07	Thompson, 2005
log Kow	4.7 - 4.8	Sietz, 2005; Ying, 2007
Log Kd*	4.3	Samsoe-Petersen, 2003
pKa 20°C	8.14	Samsoe-Petersen, 2003
Vapor Pressure (Pa) 25°C	7.00E-04	Samsoe-Petersen, 2003

*for activated sludge with organic carbon content around 40%

Behavioral Expectations:

Sorption
Sorption to the fraction of organic carbon in the soil can be expected. (Moderate)

Biodegradation
Laboratory studies have shown triclosan to be highly biodegradable, with mineralization half-lives ranging from 15 to 35 days (Ciba, 2001). However, McAvoy et al. (2002) demonstrated that triclosan is only biodegradable under aerobic conditions—no biodegradation occurred under anaerobic conditions. Ying et al. (2007) contains more ½ life data.

Abiotic degradation
--Photolysis--
Applied to waters with pHs above 8. Otherwise photolysis is negligible (Thompson et al., 2005).

Ecotoxicity:
Algae appear to be the most sensitive aquatic organism to triclosan. EC_{50} values for inhibition of growth rate have been found in the range 1.4 – 19 μg/l. The lowest reported NOEC is 0.69 μg/l (Orvos et al., 2002).

Organism		$L(E)C_{50}$	Data-source
Fish	Trout (96 h)	0.260 mg/l	Ciba
Crustacea	*C. dubia* (48 h)	0.130 mg/l	Ciba
	Daphnia magna (48 h)	0.390 mg/l	Ciba
Algae	*Scenedesmus sub.* (72 h)	0.0007 mg/l	Ciba
	Scenedesmus sub. (72h, 96 h)	0.0014 mg/l	Ciba
	Anabena flos-acuae	0.000966 mg/l	U.S. EPA
	Selenastrum capricornutum	0.00446 mg/l	U.S. EPA

Degradation Daughter Products:
Methyl-Triclosan
Dioxin (Glaser, 2004)

Sources:

Ciba Specialty Chemicals (2001). General Information on Chemical, Physical, and Microbiological Properties fo Irgasan DP 300, Irgacare MP, and Irgacide LP10. Brochure 2520. Publication AgB2520e.02.2001. Ciba Speciality Chemicals: Basel, Switzerland.

Conn, K.E., L.B. Barber, et al. (2006). "Occurrence and Fate of Organic Contaminants during Onsite Wastewater Treatment." Environmental Science and Technology 40(23): 7358-7366. a41.

DeJong, K.E., R.L. Siegrist, et al. (2004). Occurrence of emerging organic chemicals in wastewater effluents from onsite systems. National Symposium on individual and small community sewage systems Sacramento, California USA, ASAE Publication Number 701P0104.

Glaser, A. (2004). "The Ubiquitous Triclosan: A common antibacterial agent exposed." Beyond Pesticides Fact Sheet http://www.beyondpesticides.org/pesticides/factsheets/Triclosan%20cited.pdf.

Kolpin, D.W., E.T. Furlong, et al. (2001). "Pharmaceuticals, Hormones, and Other Organic Wastewater Contaminants in U.S. Streams, 1999-2000: A National Reconnaissance." Environmental Science and Technology **36**(6): 1201-1211. a7

Lindstroem, A., I.J. Buerge, et al. (2002). "Occurrence and Environmental Behavior of the Bactericide Triclosan and Its Methyl Derivative in Surface Waters and in Wastewater." Environmental Science and Technology **36**(11): 2322-2329. d5. 3004

McAvoy, D.C., A.J. DeCarvalho, et al. (2002). "Investigation of an onsite wastewater treatment system in sandy soil: modeling the fate of surfactants." Environmental Toxicology and Chemistry **21**(12): 2623-2630.

Orvos, D.R., D.J. Versteeg, et al. (2002). "Aquatic Toxicity of Triclosan." Environmental Toxicology and Chemistry **21**(7): 1338-1349.

Samsøe-Petersen, L., M. Winther-Nielsen, et al. (2003). Fate and effects of triclosan. Environmental Project No. 861. D. W. a. Environment, Danish Environmental Protection Agency.

Seitz Jr, E.P., A.L. Waggoner, et al. (2005). Compositions having enhanced deposition of a topically active compound on a surface, Google Patents.

Thompson, A., P. Griffin, et al. (2005). "The Fate and Removal of Triclosan during Wastewater Treatment." Water Environment Research **77**(1): 63-67. d8. 3007

Ying, G.G. and R.S. Kookana (2007). "Triclosan in wastewaters and biosolids from Australian wastewater treatment plants." Environmental International **33**(2): 199-205. d4. 3003

A.2 4-Nonylphenol Fact Sheet (4-NP)

Molecule Image:

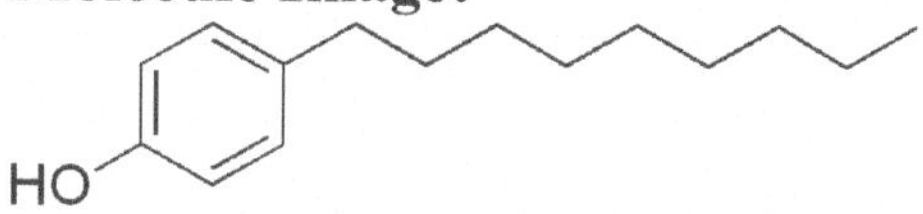

Chemical Formula:
C9H19(C6H4)OH

Uses:
Nonionic surfactant (daughter product of nonylphenol polyethoxylates) widely used as industrial detergents, emulsifiers, wetting agents, and dispersing agents

Occurrence:

Water Source	Median Concentration (µg/L)	Maximum (µg/L)	Sources
Wastewater	>18	>38	Conn, 2006
Septic Tank Effluent	4.4 res 19 non res	58 res 340 nonres	Conn, 2006
River water	0.8	40	Kolpin, 2002
Lake water	--	--	--

Chemical Properties:

Parameter	Value Range	Sources
log Koc	4 - 5.9	Ahel, 1993; McLeese, 1981; Duering, 2002
Solubility (mg/L)*	4.6 - 11.9	Ahel, 1993
Kh (atm-m3/mole)	2.45E-09	Hellmann, 1987
log Kow	4.2 - 4.48	Ahel, 1993, McLeese, 1981
Kd	80-120	Duering, 2002
pKa 20°C	10.7	McLeese, 1981
Vapor Pressure (atm) 25°C	1.00E-06	Ahel, 1993

*properties, especially solubility, are strongly pH dependent for 4-NP (solubility decreases with pH; lower end of range = pH of 5, upper end = pH of 9)

Behavioral Expectations:

Sorption
Sorption to the organic carbon fraction of the soil is expected. Removal in activated sludge is highly effective (Ahel, 1993; Ahel, 1994).

Biodegradation
Primarily aerobic though anaerobic biodegradation does occur at a much slower rate. At 16 deg C, 1.0 ppm nonylphenol is stream and pond water samples degrade with a half-life of 2.5 days if the reaction flasks were open and 16 days if they were closed (Sundaram, 1981).

Abiotic degradation
--Photolysis--

Important in lake water. In the summer at noon with clear skies the half-life measured due to photolysis is 10-15hrs (Ahel, 1994).

Ecotoxicity:
4-NP is a known endocrine disrupting compounds and is considered toxic to aquatic life. Regulation have been set by the U.S. EPA and are listed below.
The bioconcentration factor is not measured nearly as high as expected (Hellmann, 1987). Measured BCFs remained on the order of 100. Predicted ones exceed 10,000.

EPA Regulations:
--Freshwater Aquatic Life--
Freshwater aquatic life and their uses should not be affected if the one-hour average concentration of nonylphenol does not exceed 28 ug/L more than once every three years on the average (acute criterion) and if the four-day average concentration of nonylphenol does not exceed 6.6 ug/l more than once every three years on the average (chronic criterion).
--Saltwater Aquatic Life--
Saltwater aquatic life and their uses should not be affected if the one-hour average concentration of nonylphenol does not exceed 7.0 ug/L more than once every three years on the average (acute criterion) and if the four-day average concentration of nonylphenol does not exceed 1.7 ug/L more than once every three years on the average (chronic criterion).

Degradation Daughter Products:
--nonylphenol polyethoxylates: (4-nonylphenol is the daughter product)
4-nonylphenolmonoethoxylate, 4-nonylphnoldiethoxylate, 4-nonylphenoltriethoxylate, 4nonylphenoltetraethoxylate, nonylphenolmonoethoxycarboxylate, nonylphenoldiethoxycarboxylate, nonylphenoltriethoxycarboxylate, nonylphneoltretraethoxycarboxylate
--4-nonylphenol derivatives:
4-tert-octylphenol, 4-tert-octylphenolmonoethoxylate, 4-tert-octylphenoldiethoxylate, 4-tert-octylphenoltriethoxyalte, 4-tert-octylphenoltriethoxylate

Sources:

Ahel, M. and W. Giger (1993). "Aqueous solubility of alkylphenols and alkylphenol polyethoxylates." Chemosphere **26**(8): 1461-1470.

Ahel, M., W. Giger, et al. (1994). "Behaviour of alkylphenol polyethoxylate surfactants in the aquatic environment- I. Occurrence and transformation in sewage treatment." Water Research **28**(5): 1131-1142.

Buser, H.R., T. Poiger, et al. (1998). "Occurrence and fate of the pharmaceutical drug diclofenac in surface waters: Rapid photodegradation in a lake." Environmental Science and Technology **32**(22): 3449-3456. g2. 3027

Conn, K.E., L.B. Barber, et al. (2006). "Occurrence and Fate of Organic Contaminants during Onsite Wastewater Treatment." Environmental Science and Technology 40(23): 7358-7366. a41.

Duering, R. (2002). "Sorption Behavior of Nonylphenol in Terrestrial Soils." Environmental Science and Technology **36**(19): 4052-4057.

Hellmann H; Z. Fresenuis. Anal Chem 328: 475-9. 1987: http://www.nies.go.jp/edc/edcdb/HomePage_e/medb/chem/chempdf/chem69.pdf

Kolpin, D.W., E.T. Furlong, et al. (2001). "Pharmaceuticals, Hormones, and Other Organic Wastewater Contaminants in U.S. Streams, 1999-2000: A National Reconnaissance." Environmental Science and Technology **36**(6): 1201-1211. a7.

McLeese, D.W., V. Zitko, et al. (1981). "Lethality and Accumulation of Alkylphenols in Aquatic Fauna." Chemosphere **10**(7): 723-730.

Sundaram, K.M. and S. Szeto (1981). "The dissipation of nonylphenol in stream and pond water under simulated field conditions." Environmental Science and Health **16**(6): 767-76.

A.3 1,4 - Dichlorobenzene Fact Sheet

Molecule Image:

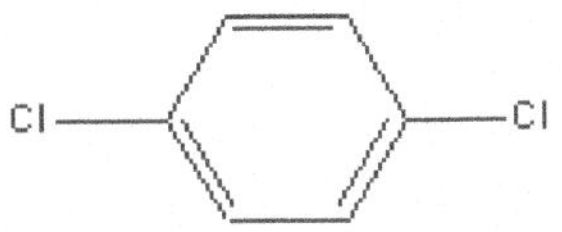

Chemical Formula:
$C_6H_4Cl_2$

Uses:
Disinfectant, deodorant, and pesticide (bug repellent)

Occurrence:

Water Source	Median Concentration (µg/L)	Maximum (µg/L)	Sources
Wastewater	<0.63	0.63	Conn, 2006
Septic Tank Effluent	1.2 res 7.0 nonres	2.1 res 59 nonres	Conn, 2006
River water	0.09	4.3	Kolpin, 2001
Lake water	--	--	

Chemical Properties:

Parameter	Value Range	Sources
log Koc	2.44	Chiou, 1983
Solubility (mg/L)	80	Yalkowsha, 2003
Kh (atm-m3/mole)	2.41E-03	Bahadur, 1997
log Kow	3.44	Hansch, 1995
Kd	--	--
pKa 20°C	--	--
Vapor Pressure (mm hg) 25°C	--	--

Behavioral Expectations:

Sorption

Sorption significantly retards 1,4-dichlorobenzene. A retardation factor of 3.4 was calculated/measure during transport through a sandy soil (Wilson et al., 1981). Based on experimental adsorption data, 1,4-dichlorobenzene can be moderately to tightly adsorbed in soil (Pelish et al., 2003).

Biodegradation

Aerobic biodegradation in water may be possible; however, anaerobic biodegradation is not expected to occur. It is possible that 1,4-dichlorobenzene will be slowly biodegraded in soil under aerobic conditions (Pelish et al., 2003).

Volatilization

Volatilization from soil surfaces may be an important transport mechanism; however, volatilization may be attenuated by tight adsorption. Volatilization was also found to be predominant elimination mechanism of 1,4-dichlorobenzene from Lake Zurich in Switzerland based on one-year monitoring studies and laboratory studies (Pelish et al., 2003).

Abiotic degradation

--Photolysis—

Chemical transformation processes such as hydrolysis, oxidation or direct photolysis (on soil surfaces) are not expected to occur (Pelish et al., 2003).

Ecotoxicity:

Possible carcinogen, neurotoxin, endocrine disruptor

Degradation Daughter Products:

Not persistent

Sources:

Bahadur, N.P., W.Y. Shiu, et al. (1997). "Temperature dependence of octanol-water partition coefficient for selected chlorobenzenes." Chemical Engineering Data **42**(4): 685–688.

Chiou et al., 1983 http://www.atsdr.cdc.gov/toxprofiles/tp10-c4.pdf

Conn, K.E., L.B. Barber, et al. (2006). "Occurrence and Fate of Organic Contaminants during Onsite Wastewater Treatment." Environmental Science and Technology 40(23): 7358-7366. a41.

Hansch et al., 1995 http://www.atsdr.cdc.gov/toxprofiles/tp10-c4.pdf

Kolpin, D.W., E.T. Furlong, et al. (2001). "Pharmaceuticals, Hormones, and Other Organic Wastewater Contaminants in U.S. Streams, 1999-2000: A National Reconnaissance." Environmental Science and Technology **36**(6): 1201-1211. a7.

Pelish, J., D. Slusher, et al. (2003). Spectrum Laboratories Chemical Fact Sheet: Compound 106467 http://www.speclab.com/index.shtml. Ft. Lauderdale, FL & Savannah, GA.

Wilson, J.T., C.G. Enfield, et al. (1981). "Transport and fate of selected organic pollutants in a sandy soil." Journal of Environmental Quality **10**(4). f2. 3038

Yalkowsky, S.H. and Y. He (2003). Handbook of Aqueous Solubility Data, CRC Press.

A.4 17β- Estradiol Fact Sheet

Molecule Image:

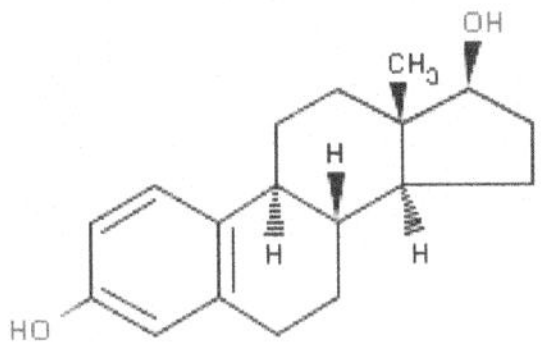

Chemical Formula:
$C_{18}H_{24}O_2$

Uses:
Used for hormone replacement therapy, blocking hormones to treat hypoestrogenism, and for hormonal contraception.

Occurrence:

Water Source	Median Concentration (µg/L)	Maximum (µg/L)	Sources
Wastewater	0.03	0.062	Leitz, 2006
Septic Tank Effluent	0.0098	0.045	Swartz, 2006
River water	0.00016	0.0002	Kolpin, 2001
Lake water	--	--	

Chemical Properties:

Parameter	Value Range	Sources
log Koc	3.14 - 4.09	Das, 2004; Yu, 2004
Solubility (mg/L)	3.1 - 13	Yu, 2004; Lee, 2003; Mansell, 2004
Kh (atm-m3/mole)	2.64E-11	25
log Kow	3.94 - 4.1	Yu, 2004; Lee, 2003; Mansell, 2004
Kd	3.56 - 83.2	Das, 2004
pKa 20°C	10.23	Yu, 2004
Vapor Pressure (mm hg) 25°C	--	--

Behavioral Expectations:

Sorption
"Adsorption is the primary mechanism for removal… attenuation increased in the presence of bioactivity, however the organic carbon composition of the wastewater had no significant effect" (Mansell and Drewes, 2004)

Biodegradation
Not significant

Volatilization

Not significant

Abiotic degradation
--Photolysis—

Ecotoxicity:
Endocrine disrupting compound

Degradation Daughter Products:
Estrone

Sources:

Das, B.S., L.S. Lee, et al. (2004). "Sorption and Degradation of Steroid Hormones in Soils during Transport: Column Studies and Model Evaluation." Environmental Science and Technology **38**(5): 1460-1470. c3. 3032

Kolpin, D.W., E.T. Furlong, et al. (2001). "Pharmaceuticals, Hormones, and Other Organic Wastewater Contaminants in U.S. Streams, 1999-2000: A National Reconnaissance." Environmental Science and Technology **36**(6): 1201-1211. a7.

Lee, L.S., T.J. Strock, et al. (2003). "Sorption and Dissipation of Testosterone, Estrogens, and Their Primary Transformation Products in Soils and Sediment." Environmental Science and Technology **37**(18): 4098-4105. c4. 3033

Lietz, A.C. and M.T. Meyer (2006). "Evaluation of Emerging Contaminants of Concern at the South District Wastewater Treatment Plant Based on Seasonal Sampling Events, Miami-Dade County, Florida, 2004." USGS Scientific Investigations Report **2006–5240**.

Mansell, J. and J.E. Drewes (2004). "Fate of Steroidal Hormones During Soil-Aquifer Treatment." Ground Water Monitoring and Remediation **24**(2). c6. 3035

Swartz, C.H., S. Reddy, et al. (2006). "Steroid Estrogens, Nonylphenol, Ethoxylate Metabolites, and Other Wastewater Contaminants in Groundwater Affected by a Residential Septic System on Cape Cod, MA." Environmental Science and Technology **40**(16): 4892-4900. c2. 3031

Yu, Z., B. Xiao, et al. (2004). "Sorption of Steroid Estrogens to Soils and Sediments." Environmental Toxicology and Chemistry **23**(3): 531-539. c5. 3030

A.5 Diclofenac Fact Sheet

Molecule Image:

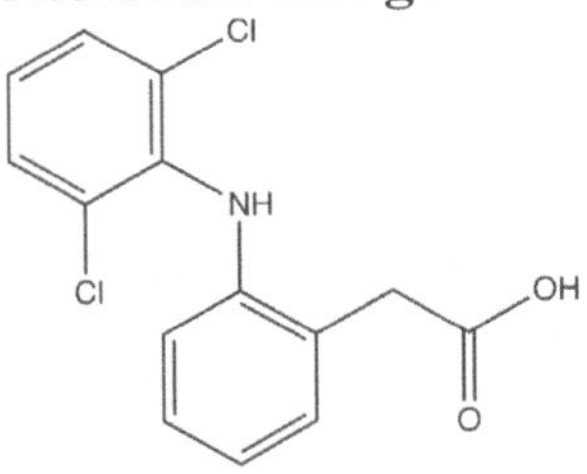

Chemical Formula:

$C_{14}H_{11}Cl_2NO_2$

Uses:

Non-steroidal anti-inflammatory drug (NSAID) taken to reduce inflammation and an analgesic reducing pain in conditions such as in arthritis or acute injury. It can also be used to reduce menstrual pain, dysmenorrhea.

Occurrence:

Water Source	Median Concentration (µg/L)	Maximum (µg/L)	Sources
Wastewater	0.72	6.2	Buser, 1998; Scheytt, 2004
Septic Tank Effluent	0.93	2.3	Buser, 1998; Scheytt, 2004
River water	<0.002	0.96	Scheytt, 2004
Lake water	--	--	

Chemical Properties:

Parameter	Value Range	Sources
log Koc	2.43 - 3.87	Beausse, 2004
Solubility (mg/L)	2.4	Scheytt, 2004
Kh (atm-m3/mole)	4.73E-12	Tixier, 2003
log Kow	4.51	Scheytt, 2004
Kd	0.8 - 5.9	Beausse, 2004
pKa 20°C	3.99 - 4.16	Scheytt, 2004; Escher, 2006
Vapor Pressure (mm hg) 25°C	6.14E-08	Beausse, 2004

Behavioral Expectations:

Sorption

Potential for sorption is moderate based on K_{oc} and K_d values. No soil transport studies have been conducted.

Biodegradation

In a biofilm reactor, very little biodegradation of diclofenac was observed with 95% of the initial concentration persisting through the unit. (Zwiener and Frimmel, 2003)

Volatilization
Due to extremely low Henry's law constant, volatization not expected to be a significant removal mechanism.

Abiotic degradation
--Photolysis—
Dominant removal mechanism

Ecotoxicity:
Causes renal failure in some organisms, most notably birds (vultures).

Degradation Daughter Products:
See Perez-Estrada et al., 2005

Sources:

Beausse, J. (2004). "Selected drugs in solid matrices: a review of environmental determination, occurrence and properties of principal substances." Trends in Analytical Chemistry **23**(10-11): 753-761. a29

Buser, H.R., T. Poiger, et al. (1998). "Occurrence and fate of the pharmaceutical drug diclofenac in surface waters: Rapid photodegradation in a lake." Environmental Science and Technology **32**(22): 3449-3456. g2. 3027

Escher, B.I., W. Pronk, et al. (2006). "Monitoring the removal efficiency of pharmaceuticals and hormones in different treatment processes of source-separated urine with bioassays." Environmental Science and Technology **40**(16): 5095-5101. a19.

Pérez Estrada, L.A., S. Malato, et al. (2005). "Photo Fenton degradation of diclofenac: Identification of main intermediates and degradation pathway." Environmental Science and Technology **39**(21): 8300-8306. g4. 3025

Scheytt, T.J., P. Mersmann, et al. (2004). "Transport of Pharmaceutically Active Compounds in Saturated Laboratory Columns." Ground Water **42**(5): 767-773. a38.

Scheytt, T.J., P. Mersmann, et al. (2006). "Mobility of pharmaceuticals carbamazepine, diclofenac, ibuprofen, and propyphenazone in miscible-displacement experiments." Journal of Contaminant Hydrology **83**(1-2): 53-69. a18.

Tixier, C., H.P. Singer, et al. (2003). "Occurrence and Fate of Carbamazepine, Clofibric Acid, Diclofenac, Ibuprofen, Ketoprofen, and Naproxen in Surface Waters." Environmental Science and Technology **37**(6): 1061-1068. G1. 3028

Zwiener, C. and F.H. Frimmel (2003). "Short-term tests with a pilot sewage plant and biofilm reactors for the biological degradation of the pharmaceutical compounds clofibric acid, ibuprofen, and diclofenac." Science of the Total Environment **309**(1-3): 201-211.

APPENDIX B

BIBLIOGRAPHY

(Additional references reviewed for the project, but not cited in the report.)

Adolfsson-Erici, M., M. Pettersson, J. Parkkonen, and J. Sturve (2002). Triclosan, a commonly used bactericide found in human milk and in the aquatic environment in Sweden. *Chemosphere* 46(9-10): 1485-1489.

Alvarez, D.A., P.E. Stackelberg, J.D. Petty, J.N. Huckins, E.T. Furlong, S.D. Zaugg, and M.T. Meyer (2005). Comparison of a novel passive sampler to standard water-column sampling for organic contaminants associated with wastewater effluents entering a New Jersey stream. *Chemosphere* 61(5): 610-622.

Anderson, D.L., A.L. Lewis, and K.M. Sherman (1989). Unsaturated zone monitoring below subsurface wastewater systems serving individual homes in Florida.

Anderson, D.L., R.J. Otis, J.I. McNeillie, and R.A. Apfel (1994). *In-situ lysimeter investigation of pollutant attenuation in the vadose zone of a fine sand.* Proceedings of the seventh National Symposium on individual and small community sewage treatment, Atlanta, GA.

Anderson, D. L., J.M. Rice, M.L. Voorhees, R.A. Kirkner, and K.M. Sherman. (1998). Ground water modeling with uncertainty analysis to assess the contamination potential from onsite sewage disposal systems (OSDS) in Florida. 264-273. St. Joseph, MI, ASAE.

Anderson, P.D., V.J. D'Aco, P. Shanahan, S.C. Chapra, M.E. Buzby, V.L. Cunningham, B.M. Duplessie, E.P. Hayes, F.J. Mastrocco, and N.J. Parke (2004). Screening analysis of human pharmaceutical compounds in U. S. surface waters. *Environmental Science and Technology* 38(3): 838-849.

Anderson, T.A., J.J. Beauchamp, and B.T. Walton (1991). Fate of volatile and semivolatile organic chemicals in soils: Abiotic versus biotic losses. *Journal of Environmental Quality* 20(2).

Andreoli, A., N. Bartilucci, R. Forgione, and R. Reynolds (1979). Nitrogen removal in a subsurface disposal system. *Journal of the American Water Pollution Control Federation* 51(4): 841-854.

Andreozzi, R., M. Raffaele, and P. Nicklas (2003). Pharmaceuticals in STP effluents and their solar photodegradation in aquatic environment. *Chemosphere* 50(10): 1319-1330.

Ashton, D., M. Hilton, and K.V. Thomas (2004). Investigating the environmental transport of human pharmaceuticals to streams in the United Kingdom. *Science of the Total Environment* 333(1-3): 167-184.

Atkins, L. and D. Christensen (2001). Alternative onsite sewage systems ability to reduce nitrogen discharged from domestic sources. Proceedings of the ninth National Symposium on individual and small community sewage treatment, Fort Worth, TX, 171-181.

Australian Soil Classification (2007). Australian Collaborative Land Evaluation Program, http://www.clw.csiro.au/aclep/asc_re_on_line/soilhome.htm.

Avery, L.M., P. Hill, K. Killham, and D.L. Jones (2004). Escherichia coli O157 survival following the surface and sub-surface application of human pathogen contaminated organic waste to soil. *Soil Biology and Biochemistry* 36(12): 2101-2103.

Ball, H.L. (1994). Nitrogen reduction in an on-site trickling filter/upflow filter wastewater treatment system. Proceedings of the seventh National Symposium on individual and small community sewage treatment, Atlanta, GA, pp. 499-503.

Bateman, E.J. and E.M. Baggs (2005). Contributions of nitrification and denitrification to N2O emissions from soils at different water-filled pore space. *Biology and Fertility of Soils*, 41: 379-388.

Beal, C.D., E.A. Gardner, and N.W. Menzies (2005). Precess, performance and pollution potential: A review of septic tank-soil absorption systems. *Australian Journal of Soil Research*, 43: 781-802.

Beauchamp, E.G., C. Gale, and J.C. Yeomans (1980). Organic matter availablity for denitrification in soils of different textures and drainage class. *Soil Science and Plant Analysis*, 11(12): 1221-1233.

Bendz, D., N.A. Paxéus, T.R. Ginn, and F.J. Loge (2005). Occurrence and fate of pharmaceutically active compounds in the environment, a case study: Höje River in Sweden. *Journal of Hazardous Materials* 122(3): 195-204.

Bester, K. (2003). Triclosan in a sewage treatment process—balances and monitoring data. *Water Research* 37(16): 3891-3896.

Blanc, R. and A. Nasser (1996). Effect of effluent quality and temperature on the persistence of viruses in soil. *Water Science and Technology* 33(10-11): 237-242.

Bombalt, C., D.C. Wolf, M.A. Gross, E.M. Rutledge, and E.E. Gbur (1994). Field performance of conventional and low pressure distribution septic systems. Proceedings of the seventh National Symposium on individual and small community sewage treatment, Atlanta, GA, 444-453.

Bound, J.P., K. Kitsou, and N. Voulvoulis (2006). Household disposal of pharmaceuticals and perception of risk to the environment. *Environmental Toxicology and Chemistry* 21(3): 301-307.

Bound, J.P. and N. Voulvoulis (2004). Pharmaceuticals in the aquatic environment – a comparison of risk assessment strategies. *Chemosphere* 56(11): 1143-1155.

Bound, J.P. and N. Voulvoulis (2006). Predicted and measured concentrations for selected pharmaceuticals in UK rivers: Implications for risk assessment. *Water Research* 40(15): 2885-2892.

Bouwer, H., (1974). Design and operation of land treatment system for minimum contamination of ground water. *Ground Water*, 12(3): 140-147.

Boyle, W.C. et al. (1994). Nitrogen removal from domestic watewater in unsewered areas. Proceedings of the seventh National Symposium on individual and small community sewage treatment", Atlanta, GA, 485-498.

Bradford, S.A., Y.F. Tadassa, and Y. Pachepsky (2006). Transport of giardia and manure suspensions in saturated porous media. *J Environ Qual* 35:749-57.

Braoudaki, M. and A.C. Hilton (2004). Low level of cross-resistance between triclosan and antibiotics in Escherichia coli K-12 and E. coli O55 compared to E. coli O157. *FEMS Microbiology Letters* 235(2): 305-309.

Brookman, S.K.E., D.R. Chadwick, and A.R. Retter (2005). Lysimeter experiments to determine the ability of soil to reduce concentrations of BOD, available P and inorganic N in dirty water. *Environmental Technology* 26(11): 1189-1204.

Brossa, L., R.M. Marcé, F. Borrull, and E. Pocurull (2005). Occurrence of twenty-six endocrine-disrupting compounds in environmental water samples from Catalonia, Spain. *Environmental Science and Technology* 24(2): 261-267.

Brusseau, M.L., J.K. Oleen, J. Santamaria, L. Cheng, P. Orosz-Coghlan, A.S. Chetochine, W.J. Blanford, P. Rykwalder, and C.P. Gerba (2005). Transport of microsporidium Encephalitozoon intestinales spores in sandy porous media. *Water Res* 39(15): 3636-42.

Bumgarner, J.R. and J.E. McCray (2007). Estimating biozone hydraulic conductivity in wastewater soil-infiltration systems using inverse numerical modeling. *Water Research* 41(11): 2349-2360.

Bunnell, J.F., R.A. Zampella, M.D. Morgan, and D.M. Gray (1999). A comparison of N removal by subsurface pressure dosing and standard septic systems in sandy soils. *J. of Environmental Management* 56: 209-219.

Campbell, S.W., M.C. Evans, and N.D. Poor (2002). Predictions of size-resolved aerosol concentrations of ammonium, chloride and nitrate at a bayside site using EQUISOLV II. *Atmospheric Environment*, 36: 4299-4307.

Carballa, M., F. Omil and J.M. Lema (2007). Calculation methods to perform mass balances of micropollutants in sewage treatment plants. Application to pharmaceutical and personal care products (PPCPs). *Environmental Science and Technology* 41(3): 884-890.

Carrera, J., I. Jubany, L. Carvallo, R. Chamy, and J. Lafuente (2004). Kinetic models for nitrification inhibition by ammonium and nitrite in a suspended and an immobilized biomass systems. *Process Biochemistry*, 39: 1159-1165.

Chattopadhyay, D., Chattopadhyay, S., Lyon W.G., and Wilson J.T. (2002). Effect of surfactants on the survival and sorption of viruses. *Environ Sci Technol* 36:19,. 4017–4024.

Cheng, L.C., A.S, I.L. Pepper, and M.L. Brusseau (2007). Influence of DOC on MS-2 bacteriophage transport in a sandy soil. *Water Air and Soil Pollution* 178: 315-322.

Chuanchuen, R., K. Beinlich, T.T. Hoang, A. Becher, R.R. Karkhoff-Schweizer, and H. Schweizer (2001). Cross-resistance between triclosan and antibiotics in Pseudomonas aeruginosa is mediated by multidrug efflux pumps: Exposure of a susceptible mutant strain to triclosan select nfxB mutants overexpressing MecxCD-OprJ. *Antimicrobial Agents and Chemotherapy* 45(2): 428-432.

Chun, S., J. Lee, R. Geyer, D.C. White, and D.R. Raman (2005). Effect of agricultural antibiotics on the persistence and transformation of 17ß-estradiol in a Sequatchie loam. *Journal of Environmental Science and Health* 40(5): 741-751.

Clara, M., B. Strenn, O. Gans, E. Martinez, N. Kreuzinger, and H. Kroiss (2005). Removal of selected pharmaceuticals, fragrances and endocrine disrupting compounds in a membrane bioreactor and conventional wastewater treatment plants. *Water Research* 39(19): 4797-4807.

Cogger, C.G. and B.L. Carlile (1984). Field performance of conventional and alternative septic systems in wet soils. *J. Environmental Quality* 13(1): 137-142.

Conn, K.E., L.B. Barber, G.K. Brown, and R.L. Siegrist (2006). Occurrence and fate of organic contaminants during onsite wastewater treatment. *Environ. Sci. Technol.*, 40(23): 7358-66.

Converse, J.C. and E.J. Tyler (1998). Soil treatment of aerobically treated domestic wastewater with emphasis on modified mounds. *In*: D.M. Sievers (Editor), Proceedings of the Eight National Symposium on Individual and Small Community Sewage Systems, Orlando, FL, 306-319.

Converse, J.C., M.E. Kean, E.J. Tyler, and J.O. Peterson (1991). Bacterial and nutrient removal in wisonsin at-grade. Proceedings of the sixth National Symposium on individual and small community sewage treatment", Chicago, IL, 46-61.

Converse, J.C., E.J. Tyler, and S.G. Litman (1994). *Nitrogen and fecal coliform removal in Wisconsin mound system*. Proceedings of the seventh National Symposium on individual and small community sewage treatment, Atlanta, GA.

Cordy, G.E., N.L. Duran, H. Bouwer, R.C. Rice, E.T. Furlong, S.D. Zuagg, M.T. Meyer, L.B. Barber, and D.W. Kolpin (2004). Do pharmaceuticals, pathogens, and other organic waste water compounds persist when waste water Is used for recharge? *Ground Water Monitoring and Remediation* 24(2): 58-69.

Das, K.C. and K. Xia (2007). Transformation of 4-nonylphenol isomers during biosolids composting. *Chemosphere* 2008(70): 761-768.

DeBorde, D.C., W.W. Woessner, B. Lauerman, and P.N. Ball (1998). Virus occurrence and transport in a school septic system and unconfined aquifer. *Ground Water* 36(5): 825-834.

Degen, M.B., R.B.J. Reneau, C. Hagedorn, and D.C. Martens (1991). Denitrification in onsite wastewater treatment and disposal systems, Virginia Water Resources Research Center Bulletin. Virginia Water Resources Research Center Virginia.

Diab, S. and M. Shilo (1988). Nitrogen transformation in wastewater reclamation. *Water Res.*, 22(5): 557-563.

Dillon, P.J. and F.H. Rigler (1974). The phosphorus-chlorophyll relationship in lakes. *Limnology and Oceanography* 19(5): 767-773.

Dix, S.P. (2001). Affects of sand texture on sidewall absorption of septic tank effluent, Proceedings of the ninth National Symposium on industrial and small community sewage systems, St. Joseph, MI, 159-170.

Dizer, H., B. Fischer, I. Sepulveda, E. Loffredo, N. Senesi, F. Santana, and P.D. Hansen (2002). Estrogenic effect of leachates and soil extracts from lysimeters spiked with sewage sludge and reference endocrine disrupters. *Environmental Toxicology and Chemistry* 17(2): 105-112.

Doi, J., K.H. Marks, A.J. DeCarvalho, D.C. McAvoy, A.M. Nielsen, L. Kravetz, and M.L. Cano (2002). Investigation of an onsite wastewater treatment system in sandy soil: sorption and biodegradation of linear alkylbenzene sulfonate. *Environmental Toxicology and Chemistry* 21(12): 2617-2622.

Drillia, P., K. Stamatelatou, and G. Lyberatos (2005). Fate and mobility of pharmaceuticals in solid matrices. *Chemosphere* 60(8): 1034-1044.

Ellis, J.B. (2006). Pharmaceutical and personal care products (PPCPs) in urban receiving waters. *Environmental Pollution* 144(1): 184-189.

Environmental Science and Engineering Inc. (1993). Onsite sewage disposal system research on the northern periphery of Lake Okeechobee. ESE No. 3-91-3010, Florida Department of Health and Rehavilitative Services.

Falconer, I.R., H.F. Chapman, M.R. Moore, and G. Ranmuthugala (2006). Endocrine-disrupting compounds: A review of their challenge to sustainable and safe water supply and water reuse. *Environmental Toxicology* 21(2): 181-191.

Gambrell, R.P., J.W. Gilliam, and S.B. Weed (1975). Nitrogen losses from soils of the North Carolina coastal plain. *J. Environmental Quality*, 4(3): 317-323.

Gilmour, C.M., F.E. Broadbent, and S.M. Beck (1977). Recycling of carbon and nitrogen through land disposal of various wastes. In: L.F. Elliot and F.J. Stevenson (Editors), Soils for Management of Organic Wastes and Waste Waters. Soil Science Society of America / American Society of Agronomy, Muscle Shoals, AL.

Ginn, T.R., B.D. Wood, K.E. Nelson, T.D. Scheibe, E.M. Murphy, and T.P. Clement (2002). Processes in microbial transport in the natural subsurface. *Advances in Water Resources* 25(8-12): 1017-1042.

Gold, A.J., W.R. Deragon, W.M. Sullivan, and J.L. Lemunyon (1990). Nitrate-nitrogen losses to groundwater from rural and suburban land uses. *J. of Soil and Water Conservation*, 45: 305-310.

Groffman, P.M. and J.M. Tiedje (1988). Denitrification hysteresis during wetting and drying cycles in soils. *Soil Science Society of America Journal*, 52(6): 1626-1629.

Groffman, P.M. and J.M. Tiedje (1991). Relationships between denitrification, CO2 production and air-filled porosity in soils of different texture and drainage. *Soil Biology & Biochemistry*, 23(3): 299-302.

Günkör, K. and K. Ünlö, (2005). Nitrite and nitrate removal efficiencies of soil aquifer treatment columns. *Turkish Journal of Engineering and Environmental Sciences*, 29: 159-170.

Haggard, B.E., J.M. Galloway, W.R. Green, and M.T. Meyer (2006). Pharmaceuticals and Other Organic Chemicals in Selected North-Central and Northwestern Arkansas Streams. *Journal of Environmental Quality* 35(4): 1078.

Halling-Sørensen, B., S. Nors Nielsen, P.F. Lanzky, F. Ingerslev, H.C. Holten Lützhøft, and S.E. Jørgensen (1998). Occurrence, fate and effects of pharmaceutical substances in the environment-A review. *Chemosphere* 36(2): 357-393.

Hari, A.C., R.A. Paruchuri, D.A. Sabatini, and T.C.G. Kibbey (2005). Effects of pH and Cationic and Nonionic Surfactants on the Adsorption of Pharmaceuticals to a Natural Aquifer Material. *Environmental Science and Technology* 39(8): 2592-2598.

Heidler, J. and R.U. Halden (2006). Mass balance assessment of triclosan removal during conventional sewage treatment. *Chemosphere* 66(2): 362-369.

Hijnen, W.A., J.F. Schijven, P. Bonne, A. Visser, and G.J. Medema (2004). Elimination of viruses, bacteria and protozoan oocysts by slow sand filtration. *Water Sci Technol* 50(1): 147-54.

Horan, N.J. (2003). Faecal Indicator Organisms. *The Handbook of Water and Wastewater Microbiology*. D. Mara and N.J. Horan. San Diego, CA, Academic Press: 104-112.

Hua, J., P. An, J. Winter, and C. Gallert (2003). Elimination of COD, microorganisms and pharmaceuticals from sewage by trickling through sandy soil below leaking sewers. *Water Research* 37(18): 4395-4404.

Hudson, B., H. Beller, C.M. Bartel, S. Kane, C. Campbell, A. Grayson, N. Liu, and S. Burastero (2005). Environmental transport and fate of endocrine disruptors from non-potable reuse of municipal wastewater, UCRL-TR-217159, Lawrence Livermore National Laboratory (LLNL), Livermore, CA.

Hutchins, S.R., M.B. Tomson, J.T. Wilson, and C.H. Ward (1984). Fate of trace organics during rapid infiltration of primary wastewater at Fort Devens, Massachusetts. *Water Research* 18(8).

Hutchins, S.R., M.B. Tomsons, P.B. Bedient, and C.H. Ward (1985). Fate of trace organics during land application of municipal wastewater. *CRC Critical Reviews in Environmental Control* 15(4): 355-417.

Ishibashi, H., N. Matsumura, M. Hirano, M. Matsuoka, H. Shiratsuchi, Y. Ishibashi, Y. Takao, and K. Arizono (2004). Effects of triclosan on the earlylife stages and reproduction of medaka Oryzias latipes and induction of hepatic vitellogenin. *Aquatic Toxicology* 67(2): 167-179.

Johnson, A.C., V. Keller, R.J. Williams, and A. Young (2007). A practical demonstration in modelling diclofenac and propranolol river water concentrations using a GIS hydrology model in a rural UK catchment. *Environmental Pollution* 146(1): 155-165.

Joss, A., S. Zabczynski, A. Göbel, B. Hoffmann, D. Löffler, C.S. McArdell, T.A. Ternes, A. Thomsen, and H. Siegrist (2006). Biological degradation of pharmaceuticals in municipal wastewater treatment: Proposing a classification scheme. *Water Research* 40(8): 1686-1696.

Kimura, K., S. Toshima, G. Amy. and Y. Watanabe (2004). Rejection of neutral endocrine disrupting compounds (EDCs) and pharmaceutical active compounds (PhACs) by RO membranes. *Journal of Membrane Science* 245(1-2): 71-78.

Kostich, M.S. and J.M. Lazorchak (2008). Risks to aquatic organisms posed by human pharmaceutical use. *Science of the Total Environment* 389(2-3): 329-339.

Kraft, G.J., B.A. Browne, W.M. DeVita, D.J. Mechenich, and M. Siddiqui (2008). Agricultural pollutant penetration and steady state in thick aquifers. *Ground Water*, 46(1): 51-60.

Krahe, S., R.A. Düring, J.A. Huisman, A.L. Horn, and S. Gäth (2006). Statistical Modeling of the Partitioning of Nonylphenol in Soil. *Water, Air, and Soil Pollution* 172(1): 221-237.

Kyveryga, P.M., A.M. Blackmer, J.W. Ellsworth, and R. Isla (2004). Soil pH effects on nitrification of fall-applied anhydrous ammonia. *Soil Science Society of Amer. Journal*, 68: 545-551.

Laak, R. (1982). On-site soil systems, N removal. *Alternative wastewater treatment*. A. S. Eikum and R. W. Seabloom, D. Reidel Publishing Co.: 129-143.

Laegdsmand, M., O.H. Jacobsen, H. Andersen, and B. Halling-Sørensen (2006). Transport and fate of steroid estrogens in slurry-amended soil. NJF373:16-22. (Available at http://chaos.bibul.slu.se/sll/njf/utredn_rapporter/NUR06-05/NUR06-05C.PDF)

Langergraber, G. (2003). Simulation of subsurface flow constructed wetlands - results and further research needs. *Water Science and Technology*, 48(5): 157-166.

Larsen, J.E., R. Langston, and G.F. Warren (1958/1951). Studies on the leaching of applied labeled phosphorus in organic soils. *Soil Science Society of America Proceedings* 22: 558-560.

Lech, J.J., S.K. Lewis, and L. Ren (1996). In Vivo Estrogenic Activity of Nonylphenol in Rainbow Trout. *Fundamental and Applied Toxicology* 30(2): 229-232.

Lee, C.J. and T.J. Rasmussen (2006). Occurrence of organic wastewater compounds in effluent-dominated streams in Northeastern Kansas. *Science of the Total Environment* 371(1-3): 258-269.

Lee, S., D.C. McAvoy, J. Szydlik, and J.L. Schnoor (1998). Modeling the fate and transport of household chemicals in septic systems. *Ground Water* 36(1): 123-132.

Lowe, K.S., S. Van Cuyk, E. Dodson, and R.L. Siegrist (2001). Field evaluation of virus treatment efficiency by wastewater soil absorption systems with aggregate-free and aggregate-laden infiltrative surfaces, Colorado School of Mines, Golden, CO.

Lowe, K.S., S.M. Van Cuyk, R.L. Siegrist, and J.E. Drewes (2008). Field Evaluation of the Performance of Engineered Onsite Wastewater Treatment Units. ASCE. *Journal of Hydrologic Engineering*.

McAvoy, D.C., C.E. White, B.L. Moore, and R.A. Rapaport (1994). Chemical fate and transport in a domestic septic system: sorption and transport of anionic and cationic surfactants. *Environmental Toxicology and Chemistry* 13(2): 213-221.

McBain, A.J., R.G. Bartolo, C.E. Catrenich, D. Charbonneau, R.G. Ledder, B.B. Price, and P. Gilbert (2003). Exposure of Sink Drain Microcosms to Triclosan: Population Dynamics and Antimicrobial Susceptibility. *Applied and Environmental Microbiology* 69(9): 5433-5442.

McCarthy, B. et al. (1998). Performance of alternative treatment systems in Northern Minnesota. *In*: D.M. Sievers (Editor), Proceedings of the Eighth National Symposium on Individual and Small Community Sewage Systems. ASAE, Orlando, FL, 446-457.

McCray, J.E., S.L. Kirkland, R.L. Siegrist, and G.D. Thyne (2005). Model parameters for simulating fate and transport of on-site wastewater nutrients. Great Plains System Research Unit, USDA-ARS, Fort Collins, CO. Technical Report No. 4.

McCray, J.E., G. Oldham, and J. Drewes (2006). Investigation of nonequilibrium physical, chemical, and microbiological transport processes for emerging organic contaminants using models and laboratory tracer tests. *Geological Society of America Abstracts with Programs* 38(7).

Miles-Richardson, S.R., V.J. Kramer, S.D. Fitzgerald, J.A. Render, B. Yamini, S.J. Barbee, and J.P. Giesy (1999). Effects of waterborne exposure of 17 ß-estradiol on secondary sex characteristics and gonads of fathead minnows (Pimephales promelas). *Aquatic Toxicology* 47(2): 129-145.

Miller, J.H., W.P. Ela, K.E. Lansey, P.L. Chipello, and R.G. Arnold (2006). Nitrogen transformations during soil-aquifer treatment of wastewater effluent-oxygen effects in field studies. *J. of Environmental Engineering*, 132(10): 1298-1306.

Mones, E. (not dated). Studies on the terrestrial toxicity of Triclosan *Manuscript or report from Ciba (available through the Danish EPA)*.

Monferran, M.V., D.A. Wunderlin, J. Nimptsch, and S. Pflugmacher (2007). Biotransformation and antioxidant response in Ceratophyllum demersum experimentally exposed to 1, 2-and 1, 4-dichlorobenzene. *Chemosphere* 68(11): 2073-2079.

Montgomery-Brown, J., J.E. Drewes, P. Fox, and M. Reinhard (2003). Behavior of Alkylphenol Polyethoxylate Metabolites During Soil Aquifer Treatment. *Water Research* 2003(37): 3673-3683.

Mote, C.R., J.W. Kleene, and J.S. Allison (1991). On-site wastewater renovation utilizing a partially saturated sand filter for nitrogen removal and a lawn sprinkler for effluent discharge. Proceedings of the sixth National Symposium on individual and small community sewage treatment, St. Joseph, MI, 173-181.

Myrold, D.D. and J.M. Tiedje (1985). Establishment of denitrification capacity in soil: effects of carbon, nitrate, and moisture. *Soil Biology & Biochemistry*, 17(6): 819-822.

Nicosia, L.A., J.B. Rose, L. Stark, and M.T. Stewart (2001). A field study of virus removal in septic tank drainfields. *J Environ Qual* 30(6): 1933-9.

Nielsen, A.M., A.J. DeCarvalho, D.C. McAvoy, L. Kravetz, M.L. Cano, and D.L. Anderson (2002). Investigation of an onsite wastewater treatment system in sandy soil: Site characterization and fate of anionic and nonionic surfactants. *Environmental Toxicology and Chemistry* 21(12): 2606-2616.

Oppel, J., G. Broll, D. Löffler, M. Meller, J. Römbke, and T. Ternes (2004). Leaching behaviour of pharmaceuticals in soil-testing-systems: a part of an environmental risk assessment for groundwater protection. *Science of the Total Environment* 328(1-3): 265-273.

Osenga, W. (2004). Antibiotics in the Environment. *Case Study Paper*, University of Washington: 1-12.

Packer, J.L., J.J. Werner, D.E. Latch, K. McNeill, and W.A. Arnold (2003). Photochemical fate of pharmaceuticals in the environment: Naproxen, diclofenac, clofibric acid, and ibuprofen. *Aquatic Sciences* 65(4): 342-351.

Pang, L., M. Close, M. Goltz, L. Sinton, H. Davies, C. Hall, and G. Stanton (2004). Estimation of septic tank setback distances based on transport of E. coli and F-RNA phages. *Environment International* 29(7): 907-921.

Paxeus, N. (2004). Removal of selected non-steroidal anti-inflammatory drugs (NSAIDs), gemfibrozil, carbamazepine, beta-blockers, trimethoprim and triclosan in conventional wastewater treatment plants in five EU countries and their discharge to the aquatic environment. *Water Science and Technology* 50(5): 253-60.

Peeples, J.A., K.M. Mancl, and D.L. Widrig (1991). An examination of the role of sand depth on the treatment efficiency of pilot scale intermittent sand filters. Proceedings of the sixth National Symposium on individual and small community sewage treatment, Chicago, IL, 114-124.

Pell, M. and F. Nyberg (1989). Infiltration of wastewater in a newly started pilot sand-filter system: III. Transformation of nitrogen. *J. of Environmental Quality*, 18: 463-467.

Peschel, J.M., P.K. Haan, and R.E. Lacey (2003). A SSURGO Pre-Processing Extension for the ArcView Soil and Water Assessment Tool. *ASAE Annual International Meeting*. Las Vegas, Nevada, ASAE.

Petrovic, M., S. Gonzalez, and D. Barceló (2003). Analysis and removal of emerging contaminants in wastewater and drinking water. *Trends in Analytical Chemistry* 22(10): 685-696.

Phillips, I.R. (2002). Phosphorus sorption and nitrogen transformation in two soils treated with piggery wastewater. *Australian Journal of Soil Research*, 40(2): 335-349.

Pietrogrande, M.C. and G. Basaglia (2007). GC-MS analytical methods for the determination of personal-care products in water matrices. *Trends in Analytical Chemistry* 26(11): 1086-1094.

Postma, F.B., A.J. Gold, and G.W. Loomis (1992). Nutrient and microbial movement from seasonally-used septic systems. *J. of Environmental Health*, 55(2): 5-10.

Powelson, D.K. and A.L. Mills (2001). Transport of Escherichia coli in sand columns with constant and changing water contents. *J Environ Qual* 30(1): 238-45.

Quintana, J.B. and T. Reemtsma (2004). Sensitive determination of acidic drugs and triclosan in surface and wastewater by ion-pair reverse-phase liquid chromatography/tandem mass spectrometry. *Rapid Communications in Mass Spectrometry* 18(7): 765-774.

Radcliffe, D.E., L.T. West, and J. Singer (2005). Gravel effect on wastewater infiltration from septic system trenches. *Soil Science Society of America Journal*, 69: 1217-1224.

Reiss, R., N.C. Mackay, J. Habig, and A. Griffin (2001). Probabilistic Ecological Risk Assessment for Triclosan. *Submitted to Environmental Toxicology and Chemistry* Sciences International, INC. Alexandria, VA.

Reneau, R.B.J. (1977). Changes in inorganic nitrogenous compounds from septic tank effluent in a soil with fluctuating water table. *J. Environmental Quality*, 6: 173-178.

Reneau, R.B.J. (1979). Changes in concentrations of selected chemical polutants in wet, tile drained soil system as influenced by disposal of septic tank effluents. *J. Environmental Quality*, 8: 189-196.

Riznyk, R.Z., J. Rockwell, L.C. Reid, and S.L. Reid (1993). Peat leachmound treatment of residential waste in subarctic Alaska. *Water, Air, Soil Pllution* 69(2): 165-177.

Roberts, P.H. and K.V. Thomas (2006). The occurrence of selected pharmaceuticals in wastewater effluent and surface waters of the lower Tyne catchment. *Science of the Total Environment* 356(1-3): 143-153.

Robertson, W.D. and J.A. Cherry (1995). In situ denitrification of septic system nitrate using reactive porous media barriers. Field Trials. *Ground Water* 33(1): 99-111.

Robertson, W.D., J.A. Cherry, and E.A. Sudicky (1991). Ground-water contamination from two small septic systems on sand aquifers. *Ground Water* 29(1): 82-92.

Robertson, W.D., N. Yeung, P.W. van Driel, and P.S. Lombardo (2005). High-permeability layers for remedation of ground water; Go wide, not deep. *Ground Water*, 43(4): 574-581.

Rule, K.L., V.R. Ebbett, and P.J. Vikesland (2005). Formation of chloroform and chlorinated organics by free-chlorine-mediated oxidation of triclosan. *Environmental Science and Technology* 39(9): 3176-3185.

Sanderson, H., D.J. Johnson, T. Reitsma, R.A. Brain, C.J. Wilson, and K.R. Solomon (2004). Ranking and prioritization of environmental risks of pharmaceuticals in surface waters. *Regulatory Toxicology and Pharmacology* 39(2): 158-183.

Sangsupan, H.A., D.E. Radcliffe, P.G. Hartel, M.B. Jenkins, W.K. Vencill, and M.L. Cabrera (2006). Sorption and Transport of 17 {beta}-Estradiol and Testosterone in Undisturbed Soil Columns. *Journal of Environmental Quality* 35(6): 2261.

Santamaria, J. and G.A. Toranzos (2003). Enteric pathogens and soil: a short review. *Int Microbiol* 6(1): 5-9.

Schaub, S.A., H.T. Bausum, and G.W. Taylor (1982). Fate of virus in wastewater applied to slow-infiltration land treatment systems. *Appl Environ Microbiol* 44(2): 383-94.

Schecher, W. (1997). Septic Tank Treatment System (SepTTS) Model for Cleaning Product Chemicals User's Manual. *Environmental Research Software* Version 1.0.

Scheytt, T., P. Mersmann, R. Lindstädt and T. Heberer (2005). Determination of sorption coefficients of pharmaceutically active substances carbamazepine, diclofenac, and ibuprofen, in sandy sediments. *Chemosphere* 60(2): 245-253.

Schipper, L.A. and M. Vojvodic-Vukovic (2001). Five years of nitrate removal, denitrification and carbon dynamics in a denitrification wall. *Water Research*, 35(14): 3473-3477.

Schmitt-Jansen, M., P. Bartels, N. Adler, and R. Altenburger (2007). Phytotoxicity assessment of diclofenac and its phototransformation products. *Analytical and Bioanalytical Chemistry* 387(4): 1389-1396.

Schwaiger, J., H. Ferling, U. Mallow, H. Wintermayr, and R.D. Negele (2004). Toxic effects of the non-steroidal anti-inflammatory drug diclofenac Part I: histopathological alterations and bioaccumulation in rainbow trout. *Aquatic Toxicology* 68(2): 141-150.

Shah, V., R. Ordonez, K. Jahan, and R.P. Ramachandran (2001). *Nonlinear parameter estimation of nonylphenol biodegradation*. IEEE International Conference, Mexico City, Mexico, Control Applications.

Sharma, H.D. and K.R. Reddy (2004). *Geoenvironmental engineering: site remediation, waste containment, and emerging waste management technologies*, John Wiley & Sons, Inc. Hoboken, NJ.

Sherman, K.M. and D.L. Anderson (1991). Evaluation of volatile organic compounds and conventional parameters from on-site sewage disposal systems in Florida. *6th National Symposium on Individual and Small Community Sewage Systems* 1991: 62-75.

Siegrist, R.L. and P.D. Jenssen (1990). Evaluation of sampling method effects on volatile organic compound measurements in contaminated soils. *Environmental Science and Technology* 24(9): 1387-1392.

Sikora, L.J. and R.B. Corey (1976). Fate of nitrogen and phosphorus insoils under septic tank disposal dields, Trans. of the ASAE, 866-875.

Soares, M.I., C. Braester, S. Belkin, and A. Abeliovich (1991). Denitrification in laboratory sand columns: carbon regime, gas accumulation and hydraulic properties. *Water Research*, 25(3): 325-332.

Sprague, L.A. and W.A. Battaglin (2005). Wastewater Chemicals in Colorado's Streams and Ground Water. *U.S. Geological Survey Fact Sheet 2004* 2004(3127): 4.

Stanford, B.D. and H.S. Weinberg (2007). Isotope dilution for quantization of steroid estrogens and nonylphenols by gas chromatography with tandem mass spectrometry in septic, soil, and groundwater matrices. *Journal of Chromatography A* 2007(1176): 26-36.

Stewart, L.W. and R.B.J. Reneau (1988). "Shallowly placed, low pressure distribution system to treat domestic waste water in soils with fluctuating high water tables". *J. Environmental Quality* 17(3): 499-504.

Swan, G., V. Naidoo, R. Cuthbert, R.E. Green, D.J. Pain, D. Swarup, V. Prakash, M. Taggart, L. Bekker, and D. Das (2006). Removing the threat of diclofenac to critically endangered Asian vultures. *Public Library of Science: Biology* 4(3): e66.

Swartz, C.H., S. Reddy, M.J. Benotti, H. Yin, L.B. Barber, B.J. Brownawell, and R.A. Rudel (2006). Steroid Estrogens, Nonylphenol, Ethoxylate Metabolites, and Other Wastewater Contaminants in Groundwater Affected by a Residential Septic System on Cape Cod, MA. *Environmental Science and Technology* 40(16): 4892-4900.

Tackett, K.N., K.S. Lowe, R.L. Siegrist, and S.M. VanCuyk (2004). Vadose Zone Treatment During Effluent Reclamation as Affected by Infiltrative Surface Architecture and Hydraulic Loading Rate. Proc. of the Tenth National Symposium on Individual and Small Community Sewage Systems, ASAE. 655-667.

Tan, Y., W.J. Bond and D.M. Griffin (1992). Transport of Bacteria during Unsteady Unsaturated Soil Water Flow. *Soil Sci Soc Am J* 56(5): 1331-1340.

Tare, V. and S.D. Bokil (1982). Wastewater treatment by soils: Role of particle-size distribution. *J. Environmental Quality* 11(4): 596-602.

Thomas, P.M. and G.D. Foster (2004). Determination of Nonsteroidal Anti-inflammatory Drugs, Caffeine, and Triclosan in Wastewater by Gas Chromatography–Mass Spectrometry. *Journal of Environmental Science and Health* 39(8): 1969-1978.

Thorsten, R. and J. Martin (2007). *Organic Pollutants in the Water Cycle: Properties, Occurrence, Analysis & Environmental Relevance of Polar Compounds*, Wiley-VCH.

Tinker, J.R. (1991). An analysis of nitrate-nitrogen in ground water beneath unsewered subdivisions. *Ground Water Monitoring & Remediation*, 11(1): 141-150.

Tixier, C., H.P. Singer, S. Canonica, and S.R. Muller (2002). Phototransformation of triclosan in surface waters: a relevant elimination process for this widely used biodice - laboratory studies, field measurements, and modeling. *Environmental Science and Technology* 36(16): 3482-3489.

Tomson, M.B., J. Dauchy, S.R. Hutchins, C. Curran, C.J. Cook, and C.H. Ward (1981). Groundwater contamination by trace level organics from a rapid infiltration site. *Water Research* 15: 1109-16.

U.S. Environmental Protection Agency (2002). *A Homeowner's Guide to Septic Systems*. EPA-832-B-02-005. Cincinnati, OH, U.S. Environmental Protection Agency.

van Voorthuizen, E.M., N.J. Ashbolt, and A.I. Schäfer (2001). Role of hydrophobic and electrostatic interactions for initial enteric virus retention by MF membranes. *Journal of Membrane Science* 194(1): 69-79.

Venhuizen, D., J.H. Wiersma, and J.A. Williams (1998). Washington Island Project: Evolution of the denitrifying sand filter concept. Proceedings of the eighth National Symposium on individual and small community sewage treatment, Orlando, FL.

Vinther, F.P. (1992). Measured and simulated denitrification activity in a cropped sandy and loamy soil. *Biology and Fertility of Soils*, 14: 43-48.

Viraraghavan, T. and R.G. Warnock (1974). Treatment efficiency of a septic tile system, Proceedings of the National Home Sewage Disposal Symposium (American Society of Agricultural Engineers), 48-57.

Viraraghavan, T. and S. Hashem (1986). Trace organics in septic tank effluent. *Water, Air, and Soil Pollution* 28(3): 299-308.

Walker, W.G., J. Bouma, D.R. Keeney, and F.R. Magdoff (1973). Nitrogen transformations during subsurface disposal of septic tank effluent in sands: I. Soil transformations. *J. Environmental Quality* 2: 475-479.

Waltman, E.L., B.J. Venables, and W.T. Waller (2006). Triclosan in a north Texas wastewater treatment plant and the influent and effluent of an experimental constructed wetland. *Environmental Toxicology and Chemistry* 25(2): 367-372.

Ward, M.H. et al., 2005. Workgroup report: Drinking-water nitrate and health - recent findings and research needs. *Environmental Health Perspective*, 113(11): 1607-1614.

Warnock, R.G. and N. Biswas (1981). Study of columnar denitrification for application in an on-site system. Proceedings of the Third National Symposium on individual and small community sewage treatment, Chicago, IL, 124-129.

Well, R. and D.D. Myrold (2002). A proposed method for measuring subsoil denitrification in situ. *Soil Science Society of America Journal*, 66: 507-518.

Wellings, F.M., A.L. Lewis, C.W. Mountain, and L.V. Pierce (1975). Demonstration of virus in groundwater after effluent discharge onto soil. *Applied Microbiology* 29(6): 751-757.

Westerman, R.L. and T.C. Tucker (1978). Factors affecting denitrification in a Sonoran desert soil. *Soil Science Society of America Journal*, 42(4): 596-599.

Whitmyer, R.W., R.A. Apfel, R.J. Otis, and R.L. Meyer (1991). Overview of individual onsite nitrogen removal systems. Proceedings of the sixth National Symposium on individual and small community sewage treatment, Chicago, IL, 143-154.

Wilhelm, S.R., S.L. Schiff, and W.D. Robertson (1994). Chemical fate and transport in a domestic septic system: unsaturated and saturated zone geochemistry. *Environmental Toxicology and Chemistry*, 13(2): 193-203.

Willman, B.P., G.W. Petersen, and D.D. Fritton (1981). Renovation of septic tank effluent in sand-clay mixtures. *J. of Environmental Quality*, 10: 439-444.

Winkler, E.S. and P.L.M. Veneman (1991). *A denitrification system for septic tank effluent using sphagnum peat moss*. "Proceedings of the sixth National Symposium on individual and small community sewage treatment, Chicago, IL".

Witz, R.L. (1974). Twenty-five years with the Nodak waste disposal system. Proceedings of the National Home Sewage Disposal Symposium, St. Joseph, MI, pp. 168-174.

Wlodarczyk, T. and U. Kotowska (2006). Nitrate and ammonium transformation and redox potential changes in organic soil (Eutric Histosol) treated with municipal waste water. *International Agrophysics*, 20: 69-76.

Wu, S.C. and P.M. Gschwend (1986). Sorption kinetics of hydrophobic organic compounds to natural sediments and soils. *Environmental Science and Technology* 20(7): 717-725.

Yamaguchi, T., P. Moldrup, D.E. Rolston, S. Ito, and S. Teranishi (1996). Nitrification in porous media during rapid, unsaturated water flow. *Water Res.*, 30(3): 531-540.

REFERENCES

Abit, S.M., A. Amoozegar, M.J. Vepraskas, and C.P. Niewoehner (2008). Solute transport and the capillary fringe and shallow groundwater: field evaluation. *Vadose Zone Journal* 7: 890-898.

Addiscot, T.M. and R.J. Wagenet (1985). Concepts of solute leaching in soils: a review of modeling approaches. *Journal of Soil Science* 36: 411-424.

Agency for Toxic Substances and Disease Registry (ATSDR) (1998). *Toxicological Profile for 1,4-Dichlorobenzene (Update)*. Public Health Service, U.S. Department of Health and Human Services, Atlanta, GA.

Aguirre, J., R. Bennet, G. Cripe, D. Eignor, J. Holmes, L. Huff, M. Kostich, D. Lattier, G.J. Smith, L. Touart, and J. Tietge (2008). Aquatic life criteria for contaminants of emerging concern part 1: General challenges and recommendations. *OW/ORD Emerging Contaminants Workgroup*.

Ahel, M. and W. Giger (1993). Aqueous solubility of alkylphenols and alkylphenol polyethoxylates. *Chemosphere* 26(8): 1461-1470.

Ahel, M., W. Giger and M. K. W. Research (1994). Behaviour of alkylphenol polyethoxylate surfactants in the aquatic environment- I. Occurrence and transformation in sewage treatment. *Water Research* 28(5): 1131-1142.

Ahuja, L. R., K. W. Rojas, J. D. Hanson, M. J. Shafer and L. Ma (2000). Root Zone Water Quality Model. Modeling management effects on water quality and crop production. *Water Resources Publication* Highlands Ranch, CO.

Altier, L.S., R. Lowrance, R.G. Williams, S.P. Inamdar, D.D. Bosch, J.M. Sheridan, R.K. Hubbard and D.L. Thomas (2002). Riparian ecosystem management model. USDA Agricultural Research Service, National Technical Information Service, Springfield VA. *Conservation Research Report* 46.

Amador, J.A., D.A. Potts, E.L. Patenaude, and J.H. Gorres (2008). Effects of sand depth on domestic wastewater renovation in intermittently aerated leachfield mesocosms. *Journal of Hydrologic Engineering* 13(8).

Amador, J.A., D.A. Potts, M.C. Savin, P. Tomlinson, J.H. Gorres, and E.L. Nicosia (2006). Mesocosm-scale evaluation of faunal and microbial communities of aerated and unaerated leachfield soil. *J Environ Qual* 35(4): 1160-9.

Ambus, P. and R. Lowrance (1991). Comparison of denitrification in two riparian soils. *Soil Science Society of America Journal* 55: 994:997.

Arah (1990). Modeling spatial and temporal variability of denitrification. *Biology and Fertility of Soils* 9: 71-77.

Atoyan, J.A., E.L. Patenaude, D.A. Potts, and J.A. Amador (2007). Effects of tetracycline on antibiotic resistance and removal of fecal indicator bacteria in aerated and unaerated leachfield mesocosms. *J Environ Sci Health A Tox Hazard Subst Environ Eng* 42(11): 1571-8.

Ausland, G., T.K. Stevik, J.F. Hanssen, J.C. Kohler, and P.D. Jenssen (2002). Intermittent filtration of wastewater--removal of fecal coliforms and fecal streptococci. *Water Research* 36(14): 3507-3516.

Azadpour-Keeley, A., B.R. Faulkner, and J.S. Chen (2003). Movement and longevity of viruses in the subsurface. In: Ground Water Issue EPA/540/S-03/500. 24 pgs., United States Environmental Protection Agency - Nat. Risk Management Research Laboratory, Cincinnati, OH 45268.

Bahadur, N.P., W.Y. Shiu, et al. (1997). "Temperature dependence of octanol-water partition coefficient for selected chlorobenzenes." *Chemical Engineering Data* 42(4): 685–688.

Bales, R.C., S.R. Hinkle, T.W. Kroeger, K. Stocking, and C.P. Gerba (1991). Bacteriophage adsorption during transport through porous media: chemical perturbations and reversibility. *Environ. Sci. Technol.* 25(12): 2088-2095.

Balmer, M.E., T. Poiger, C. Droz, K. Romanin, P.A. Bergqvist, M.D. Mueller, and H.R. Buser (2004). Occurrence of methyl triclosan, a transformation product of the bactericide triclosan, in fish from various lakes in Switzerland. *Environmental Science and Technology* 38(2): 390-395.

Baveye, P. and A. Valocchi (1989). An Evaluation of Mathematical Models of the Transport of Biologically Reacting Solutes in Saturated Soils and Aquifers, *Water Resour. Res.*, 25(6), 1413-1421.

Beal, C.D., E.A. Gardner, and N.W. Menzies (2005). Process, performance, and pollution potential: a review of septic tank-soil absorption systems. *Australian Journal of Soil Research* 43: 781-802.

Beare, M.H., R. Lowrance, and J.L. Meyer (1994). Biotic regulation of nitrate depletion in a Coastal Plain riparian forest: Experimental approach and preliminary results. *397 Riparian ecosystems in the humid U.S. Nat. Assoc. Conserv. Districts* Washington, DC: 388-397.

Bedessem, M.J., T.V. Edgar, and R. Roll (2005). Nitrogen removal in laboratory model leachfields with organic-rich layers. *J. Environmental Quality* 34(3): 936-942.

Beggs, R.A., G. Tchobanoglous, D. Hills, and R.W. Crites (2004). Modeling subsurface drip application of on-site wastewater treatment system effluent. Proceedings of the tenth National Symposium on individual and small community sewage treatment, Sacremento, CA.

Bellier, N., F. Chazarenc, and Y. Comeau (2006). Phosphorus removal from wastewater by mineral apatite. *Water Research* 40(15): 2965-2971.

Berger, P. (1994). Regulation related to groundwater contamination: The Draft Groundwater Disinfection Rule. In: Uri Zoller (ed.), *Groundwater Contamination and Control*, Marcel Dekker Inc., NY.

Berstrom, D.W. and E.G. Beauchamp (1993). An empirical model of denitrification. *Canada Journal of Soil Science* 73: 421-431.

Bester, K. (2005). Fate of triclosan and triclosan-methyl in sewage treatment plants and surface waters. *Environmental Contamination and Toxicology* 49(1): 9-17.

Beausse, J. (2004). Selected drugs in solid matrices: a review of environmental determination, occurrence and properties of principal substances. *Trends in Analytical Chemistry* 23(10-11): 753-761.

Bhargava, D.S. and S.B. Sheldarkar (1993). Use of TNSAC inphosphate adsorption studies and relationships. Literature, experimental methodology, justification and effects of process variables. *Water Research* 27(2): 303-312.

Bhattacharjee, S., J.N. Ryan, and M. Elimelech (2002). Virus transport in physically and geochemically heterogeneous subsurface porous media. *J Contam Hydrol* 57: 161–187.

Bingner, R.L. and F.D. Theurer (2008). AnnAGNPS Technical Processes: Documentation Version 3.2.. *Available at www.ars.usda.gov/Research/docs.htm?docid=5199* Verified 28 April, 2008.

Bitton, G. (1980). *Introduction to Environmental Virology*. New York, NY, John Wiley & Sons.

Blair, T.C. and J.G. McPherson (1999). Grain-size and textural classification of coarse sedimentary particles. *J. of Sedimentary Research* 69(1): 6-19.

Boeije, G. M., J. O. Wagner, F. Koormann, P. A. Vanrolleghem, D. R. Schowanek and T. C. J. Feijtel (2000). New PEC definitions for river basins applicable to GIS-based environmental exposure assessment. *Chemosphere* 40(3): 255-265.

Böhmer, W., H. Rüdel, A. Wenzel, and C. Schröter-Kermani (2004). Retrospective monitoring of triclosan and methyl-triclosan in fish: results from the German Environmental Specimen Bank. *Oganohalogen Compounds* 66: 1516.

Bohrer, R. (2000). Drip distribution soil performance and operations in a northern climate. MS Thesis. University of Wisconsin, Madison.

Bohrer, R. and J.C. Converse (2001). *Soil treatment performance and cold weather operation of drip distribution system*. Proceedings of the ninth National Symposium on individual and small community sewage treatment, Fort Worth, TX, ASAE.

Borden, R., and P. Bedient (1986. Transport of Dissolved Hydrocarbons Influenced by Oxygen-Limited Biodegradation 1. Theoretical Development, *Water Resour. Res.*, 22(13), 1973-1982.

Borrero, C., F. Pena, and J. Torrent (1988). Phosphate sorption by calcium carbonate in some soils of the Mediterranean part of Spain. *Geoderma* 42: 261-269.

Bouraoui, F. and T.A. Dillaha (1996). ANSWERS-2000: runoff and sediment transport model. *Journal of Environmental Engineering* 122: 493-502.

Bouwer, E.J. and G.D. Cobb (1987). Modeling of biological processes in the subsurface. *Water. Sci. Technol.*, 19, 769-779.

Bouwer, E.J., and P.L. McCarty (1984). Modeling of trace organic biotransformation in the subsurface. *Ground Water*, 22, 433-440.

Bowden, K. and L. C. Brown (1984). Relating effluent control parameters ot river quality objectives using a generalized catchment simulation-model. *Water Science and Technology* 16(5): 197-206.

Bradford, S.A., Y.F. Tadassa, and Y. Pachepsky (2006). Transport of giardia and manure suspensions in saturated porous media. *J Environ Qual* 35(3): 749-57.

Brady, N.C. and R.R. Weil (2002). *The Nature and Properties of Soil*. Upper Saddle River, NJ, Prentice Hall.

Brady, N.C. and R.R. Weil (2008). *The nature and properties of soil. 14th Edition*. Upper Saddle River, NJ, Prentice Hall.

Brandes, M., N.A. Chowhdry, and W.W. Cheng (1974). *Experimental study on removal of pollutants from domestic sewage by underdrained soil filters*. Proceedings of the national home sewage disposal dymposium, Chicago, IL.

Breidenbach, P., Chattopadhyay, S., Lyon, W.G. (2009). Survival and Transport of Viruses in the Subsurface, An Environmental Handbook, U.S. EPA Document in preparation.

Brown, A.R., A.M. Riddle, N.L. Cunningham, T.J. Kedwards, N. Shillabeer, and T.H. Hutchinson (2003). Predicting the effects of endocrine disrupting chemicals on fish populations. *Human and Ecological Risk Assessment* 9(3): 761-788.

Buser, H.R., T. Poiger and M.D. Mueller (1998). Occurrence and fate of the pharmaceutical drug diclofenac in surface waters: Rapid photodegradation in a lake. *Environmental Science and Technology* 32(22): 3449-3456.

Canter, L.W. and R.C. Knox (1985). *Septic tank system effects on ground water quality*. Chelsea, Michigan, Lewis Publishers, Inc.

Carballa, M., F. Omil, J.M. Lema, M. Llompart, C. Garci´a-Jares, I. Rodri´guez, M. Gómez, and T. Ternes (2004). Behavior of pharmaceuticals, cosmetics and hormones in a sewage treatment plant. *Water Research* 38(12): 2918-2926.

Castiglioni, S., R. Bagnati, R. Fanelli, F. Pomati, D. Calamari, and E. Zuccato (2006). Removal of pharmaceuticals in sewage treatment plants in Italy. *Environmental Science and Technology* 40(1): 357-363.

Chatenet, B., B. Marticorena, L. Gomes, and G. Bergametti (1996). Assessing the microped size distribution of desert soils erodible by wind. *Sedimentology* 43(5): 901-911.

Chattopadhyay, D., S. Chattopadhyay, W.G. Lyon, and J.T. Wilson (2002). Effect of surfactants on the survival and sorption of viruses. *Environ Sci Technol* 36(19): 4017-4024.

Chen, J.S., R.S. Mansell, P. Nkedi-Kizza, and B.A. Burgoa (1996). Phosphorus transport during transient unsaturated water flow in an acid sandy soil. *Soil Science Society of America Journal* 60: 42-48.

Chiou et al., 1983 http://www.atsdr.cdc.gov/toxprofiles/tp10-c4.pdf

Chu, Y., Jin, Y, Flury, M., Yates, M.V. (2001). Mechanisms of virus removal during transport in unsaturated porous media. *Water Resour. Res.* 37(2):253-263.

Chu, Y., Y. Jin, T. Baumann, and M.V. Yates (2003). Effect of soil properties on saturated and unsaturated virus transport through columns. *J Environ Qual* 32(6): 2017-25.

Clapp, R.B. and G.M. Hornberger (1978). Empirical equations for some soil hydraulic properties. *Water Resources Research* 14(601-604).

Codd, G.A. (1995). Cyanobacterial toxins: occurrence, properties and biological significance. *Water Science and Technology* 32(4): 149-156.

Cogger, C.G., L.M. Hajjar, C.L. Moe, and M.D. Sobsey (1988). Septic system performances on a coastal barrier island. *J. Environmental Quality* 17(3): 401-407.

Conn, K.E. (2008). Occurrence and Fate of Endocrine-Disrupting Compounds and Other Trace Organic in Onsite Wastewater Treatment Systems. Ph.D. Dissertation. Environmental Science and Engineering, CSM, Golden, CO.

Conn, K.E., L.B. Barber, G.K. Brown, and R.L. Siegrist (2006). Occurrence and fate of organic contaminants during onsite wastewater treatment. *Environmental Science and Technology* 40(23): 7358-7366.

Coogan, M.A., R.E. Edziyie, T.W. La Point, and B.J. Venables (2007). Algal bioaccumulation of triclocarban, triclosan, and methyl-triclosan in a North Texas wastewater treatment plant receiving stream. *Chemosphere* 67(10): 1911-1918.

Corapcioglu, M.Y. and A. Haridas (1984). Transport and fate of microorganisms in porous media: A theoretical investigation. *J. Hydrol.* 72:149-169.

Craun, G.F. (1985). Summary of waterborne illness transmitted through contaminated groundwater. *J. Environmental Health*. 48:122-127.

Crites, R.W. (1985). Micropollutant removal in rapid infiltration. *Artificial Recharge of Groundwater*. T. Ashano. Boston, Butterworth Publishers: 579-608.

Crites, R.W. and G. Tchobanoglous (1998). *Small and Decentralized Wastewater Management Systems*. Boston, MA, McGraw-Hill.

Crockett, C. P., R. W. Crabtree and I. D. Cluckie (1989). River quality models for consent setting in England and Walkes. *Water Science and Technology* 21(8): 1015-1024.

Darnault, C.J., P. Garnier, Y.J. Kim, K.L. Oveson, T.S. Steenhuis, J.Y. Parlange, M. Jenkins, W. C. Ghiorse, and P. Baveye (2003). Preferential transport of Cryptosporidium parvum oocysts in variably saturated subsurface environments. *Water Environ Res* 75(2): 113-20.

Das, B.S., L.S. Lee, P.S.C. Rao, and R.P. Hultgren (2004). Sorption and degradation of steroid hormones in soils during transport: Column studies and model evaluation. *Environmental Science and Technology* 38(5): 1460-1470.

Daughton, C.G. (2004). *PPCPs in the Environment: Future Research -- Beginning with the End Always in Mind*, Springer. a54

Daughton, C. and T. Ternes (1999). Pharmaceuticals and Personal Care Products in the Environment: Agents of Subtle Change? *Environ. Health Perspec*. 107(Supplement 6): 907-938.

de Camargo, O.A., J.W. Biggar, and D.R. Nielsen (1979). Transport of inorganic phosphorus in an alfisol. *Soil Science Society of America Journal* 43: 884-890.

DeBorde, D.C., W.W. Woessner, Q.T. Kiley, and P. Ball (1999). Rapid transport of viruses in a floodplain aquifer. *Water Res.* 33:10, 2229–2238.

DeJong, K.E., R.L. Siegrist, L.B. Barber, and A.L. Wren (2004). *Occurrence of emerging organic chemicals in wastewater effluents from onsite systems*. National Symposium on individual and small community sewage systems Sacramento, California USA, ASAE Publication Number 701P0104.

DHI (2003). Mike-11: a modelling system for rivers and channels, reference manual. *DHI--Water and Development*.

Discover (2002). Top 100 Science Stories of 2002. *Discover Magazine*.

Dowd, S.E., S.D. Pillai, S. Wang, and M.Y. Corapcioglu (1998). Delineating the specific influence of virus isoelectric point and size on virus adsorption and transport through sandy soils. *Appl Environ Microbiol* 64(2): 405-10.

Doyle, S. (2006). Phosphorus Transport and Fate During Onsite Wastewater Treatment. M.S. Thesis. Environmental Science & Engineering Division, Colorado School of Mines (CSM), Golden, CO.

Doyle, S., J.E. McCray, K.S. Lowe, and G.D. Thyne (2005). Modeling Phosphorus Reaction and Transport at an Experimental Onsite Wastewater Site. Proc. of the NOWRA 2005 Annual Technical Education Conference.

Drewes, J.E., T. Heberer, T. Rauch, and K. Reddersen (2003). Fate of pharmaceuticals during ground water recharge. *Ground Water Monitoring and Remediation* 23(3): 64-72.

Drewes, J.E., J. Hemming, S.J. Ladenburger, J. Schauer, and W. Sonzogni (2005). An assessment of endocrine disrupting activity changes during wastewater treatment through the use of bioassays and chemical measurements. *Water Environment Research* 77(1): 12.

Duering, R. (2002). Sorption behavior of nonylphenol in terrestrial soils. *Environmental Science and Technology* 36(19): 4052-4057.

Duncan, C., R. Reneau, Jr., and C. Hagedorn (1994). Impact of effluent quality and soil depth on renovation of domestic wastewater. *Seventh International Symposium on Individual and Small Community Sewage Systems*. Atlanta, Georgia, ASAE.

Dyer, N. (2008). Tainted Tap Water. *Popular Science* 3-10-2008(www.popsci.com/environment/article/2008-03/.).

Eichner, E. M., T. C. Cambareri, and A. J. Carbonell (1992). Cape Cod Commission, Barnstable, MA. Technical Bulletin 91-001.

Enfield, C.G., T. Phan, D.M. Walters, and R.J. Ellis (1981). Kinetic model for phosphate transport and transformation in calcareous soils: I. Kinetics of transformation. *Soil Science Society of America Journal* 45: 1059-1064.

Escher, B.I., W. Pronk, M.J.F. Suter, and M. Maurer (2006). Monitoring the removal efficiency of pharmaceuticals and hormones in different treatment processes of source-separated urine with bioassays. *Environmental Science and Technology* 40(16): 5095-5101.

Faulkner, B.R., W.G. Lyon, F.A. Khan, and S. Chattopadhyay (2002). Predicting Attenuation of Viruses During Percolation in Soils: 1. Probabilistic Model. EPA/600/R-02/051a. 43 pgs. United States Environmental Protection Agency - Nat. Risk Management Research Laboratory, Cincinnati, OH 45268.

Faulkner, B.R., W.G. Lyon, F.A. Khan, and S. Chattopadhyay (2003). Modeling leaching of viruses by the Monte Carlo method. *Water Res* 37(19): 4719-4729.

Feachem, R.G., D.J. Bradley, H. Garelick, and D.D. Mara (1980). *Health aspects of excreta and sullage management - a state-of-the-art review*. Washington, D.C., The World Bank.

Feachem, R.G., D.J. Bradley, H. Garelick, and D.D. Mara (1983). *Sanitation and Disease Health Aspects of Excreta and Wastewater Management*. Washington, D.C., The World Bank.

Field, J.P., K.L. Farrell-Poe, and J.L. Walworth (2007). Comparative treatment effectiveness of conventional trench and seepage pit systems. *Water Environ Res* 79(3): 310-9.

Fischer, E.A. (1999). Nutrient transformation and fate during intermittent sand filtration of wastewater. *ESE*. Golden, CO, Colorado School of Mines. MS Thesis.

Folorunso, O.A., and D.E. Rolston (1984). Spatial variability of field-measured denitrification gas flux. *Soil Sci. Soc. Am. J.* 48:1214-1219.

Fox, K. K., M. Daniel, G. Morris and M. S. Holt (2000). The use of measured boron concentration data from the GREAT-ER UK validation study (1996-1998) to generate predicted regional boron concentrations. *Science of the Total Environment* 2000(251): 305-316.

Frimpter, M.H., J.J. Donohue IV, and M.V. Rapacz (1990). A mass-balance model for predicting nitrate in ground water. *Journal of the New England Water Works Association* CIV: 219-232.

Fryzuk, M.D. (2004). Ammonia transformed. *Nature* 427: 498-499.

Gabrielle, B., P. Laville, C. Henault, B. Nicoullaud, and J.C. Germon (2008). Unpublished work.

Gerba, C.P. and S.M. Goyal (1985). Pathogen removal from wastewater during groundwater recharge. *Artificial Recharge of Groundwater*. T. Asano. Stoneham, MA, Butterworth Publishers: 283-317.

Gerba, C.P., C. Wallis, and J.L. Melnick (1975). Fate of wastewater bacteria and viruses in soil. *Journal of the Irrigation and Drainage Division* 101(IR3): 157-174.

Gerritse, R.G. (1993). Prediction of travel times of phosphate in soils at a disposal site for wastewater. *Water Research* 27(2): 263-267.

Gerritse, R.G., J.A. Adeney, and J. Hosking (1995). Nitrogen losses from a domestic septic tank system on the Darling Plateau in Western Australia. *Water Research* 29(9): 2055-2058.

Gill, L.W., C. O'Sulleabhain, B.D.R. Misstear, and P.J. Johnston (2007). The Treatment Performance of Different Subsoils in Ireland Receiving On-Site Wastewater Effluent. *J Environ Qual* 36(6): 1843-1855.

Glaser, A. (2004). The Ubiquitous Triclosan: A common antibacterial agent exposed. *Beyond Pesticides Fact Sheet* http://www.beyondpesticides.org/pesticides/factsheets/Triclosan%20cited.pdf.

Gold, A.J., P.A. Jacinthe, P.M. Groffman, W.R. Wright, and R.H. Puffer (1998). Patchiness in groundwater nitrate removal in a riparian forest. *Journal of Environmental Quality* 27: 146-155.

Goldscheider, N., L. Haller, J. Pote, W. Wildi, and J. Zopfi (2007). Characterizing water circulation and contaminant transport in Lake Geneva using bacteriophage tracer experiments and limnological methods. *Environ Sci Technol* 41(15): 5252-8.

Goyal, S.M. and C.P. Gerba (1979). Comparative adsorption of human enteroviruses, simian rotavirus, and selected bacteriophages to soils. *Appl Environ Microbiol* 38(2): 241-7.

Grant, R.F. and D.J. Heaney (1997). Inorganic phosphorus transformation and transport in soils: Mathematical modeling. *Ecosystems* 61: 752-754.

Groffman, P.M. and J.M. Tiedje (1991). Relationships between denitrification, CO2 production and air-filled porosity in soils of different texture and drainage. *Soil Biology and Biochemistry* 23: 299-302.

Groffman, P. M., A. J. Gold, and R. C. Simmons (1992). Nitrate dynamics in riparian forests: Microbial studies. *Journal of Environmental Quality* 21:666-671.

Grosse, W., F. Wissing, R. Perfler, Z. Wu, J. Chang, and Z. Lei (1999). Biotechnological approach to water quality improvement in tropical and subtropical areas for reuse and rehabilitation of aquatic ecosystems. Final report, INCO-DC Project Contract No. ERBIC18C18CT960059: Cologne, Germany.

Gruesbeck, C., and R.E. Collins (1982). Entertainment and deposition of fine particles in porous medis. *Soc. Petrol. Eng. J.*, 22:847-856.

Güler, C., G.D. Thyne, J.E. McCray, and A.K. Turner (2002). Evaluation of graphical and multivariate statistical methods for classification of water chemistry data. *Hydrogeology Journal* 10(4): 455-474.

Hagedorn, C. (1984). Microbiological aspects of groundwater pollution due to septic tanks. *Groundwater Pollution Microbiology*. G. Bitton and C. P. Gerba. New York, John Wiley & Sons: 181-195.

Hain, K.E. and R.T. O'Brien (1979). The survival of enteric viruses in septic tanks and septic tank drain fields. New Mexico State University, Las Cruces, NM, New Mexico Water Resources Research Institute.

Han, J., Y. Jin and C.S. Willson (2006). Virus Retention and Transport in Chemically Heterogeneous Porous Media under Saturated and Unsaturated Flow Conditions. *Environ. Sci. Technol.* 40(5): 1547-1555.

Hansch et al., 1995 http://www.atsdr.cdc.gov/toxprofiles/tp10-c4.pdf

Hanson, G. C., P. M. Groffman, and A. J. Gold (1994). Symptoms of nitrogen saturation in a riparian wetland. *Ecological Applications* 4:750-756.

Hanson, B.R., J. Šimůnek, and J.W. Hopmans (2006). Evaluation of urea-ammonium-nitrate fertigation with drip irrigation using numerical modeling. *Agricultural Water Management* 86: 102-113.

Hantzsche, N.M. and E.J. Finnemore (1992). Predicting groundwater nitrate-nitrogen impacts. *Ground Water* 30:490-499.

Harrison, R.B., N.S. Turner, J.A. Hoyle, J. Krejsl, D.D. Tone, C.L. Henry, P.J. Isaksen, and D. Xue (2000). Treatment of septic effluent for fecal coliform and N in coarse-textured soils: use of soil-only and sand filter systems. *Water Air and Soil Pollution* 124: 205-215.

Harter, T., S. Wagner and E.R. Atwill (2000). Colloid Transport and Filtration of Cryptosporidium parvum in Sandy Soils and Aquifer Sediments. *Environ. Sci. Technol.* 34(1): 62-70.

Hassan, G., R.B. Jr. Reneau, and C. Hagedorn (2008). Modeling effluent distribution and nitrate transport the on-site wastewater system. *Journal of Environmental Quality* 37:1937-1948.

Haycock, N.E., G. Pinav, and C. Walker (1993). Nitrogen retention in river corridors: European perspective. *Ambio* 22: 340-346.

Heatwole, K.K. and J.E. McCray (2007). Modeling potential vadose-zone transport of N from onsite wastewater systems at the development scale. *Journal of Contaminant Hydrology* 91: 184-201.

Heberer, T. (2002). Occurrence, fate, and removal of pharmaceutical residues in the aquatic environment: a review of recent research data. *Toxicology Letters* 131(1-2): 5-17.

Heinen, M. (2003). A simple denitrifcation model? Literature review, sensitivity analysis, and application. Alterra, Green World Research, Wageningen, The Netherlands(Alterra-rapport 690).

Henault, C., X. Devis, J.L. Lucas, and J.C. Germon (1998). Influence of different agricultural practices (type of crop, for of N-fertilizer) on soil nitrous oxide emissions. *Biology and Fertility of Soils* 27: 299-306.

Henault, C. and J.C. Germon (2000). NEMIS, a predictive model of denitrification on the field scale. *European Journal of Soil Science* 51: 257-270.

Hinkle, S. T., R. J. Weick, J. M. Johnson, J. D. Cahill, S. G. Smith and B. J. Rich (2005). Organic wastewater compounds, pharmaceuticals, and coliphage in ground water receiving discharge from onsite wastewater treatment systems near La Pine, Oregon: Occurrence and implications for transport. *U.S. Geological Survey Scientific Investigations Report 2005 5055: 98.*

Hirsch, R. M. (2008). Statement of Dr. Robert M. Hirsch, Associate Director For Water, U.S. Geological Survey. U.S. Department of the Interior Before the Committee on Environmenta and Public Works Subcommittee on Transportation Saftey, Infrastructure Security and Water Quality. April 15, 2008.

Hisashi, Y., K. Mitsu, S. Kazuo, and H. Masakazu (1986). A fundamental research on phosphate removal by using slag. *Water Research* 20(5): 547-557.

Horan, N. (2003). Protozoa. *The Handbook of Water and Wastewater Microbiology*. D. Mara and N. J. Horan. San Diego, CA, Academic Press: 69-76.

Hseu, Z.Y. (2006). Response of microbial activities in two contrasting soils to 4-nonylphenol treated with biosolids. *Chemosphere* 64(10): 1769-1776.

Hsu, B.M., C. Huang, and J.R. Pan (2001). Filtration behaviors of Giardia and Cryptosporidium--ionic strength and pH effects. *Water Res* 35(16): 3777-82.

Hua, W., E.R. Bennett, and R.J. Letcher (2005). Triclosan in waste and surface waters from the upper Detroit River by liquid chromatography-electrospray-tandem quadrupole mass spectrometry. *Environmental International* 31(5): 621-630.

Hummel, P.R., J.L. Kittle, P.B. Duda, and A.S. Patwardhan (2003). *Calibration of a Watershed Model for Metropolitan Atlanta*. WEF Specialty Conference Proceedings, Chicago, Illinois.

Hurst, C.J. (1991). Modeling the environmental fate of microorganisms. *Americ. Soc. Mircrobio.*, Washington DC, 292 p.

Hutson, J.L. (2003). LEACHM: Model description and user's guide. School of chemistry, physics, and earth sciences. The Flinders Univ. of South Australia, Adelaide

Hutson, J.L. and R.J. Wagenet (1992). LEACHM - Leaching estimation and chemistry model. Version 3., Res. Ser. 92-3. Department of Soil, Crop, and Atmospheric Science, Cornell University, Ithica, NY.

Jabro, J.D., J.D. Toth, X. Dou, R.H. Fox, and D.D. Fritton (1995). Evaluation of N version of LEACHM for predicting nitrate leaching. *Soil Science* 160: 209-217.

Jang, L.K., P.W. Chang, J.E. Findley, and T.F. Yen (1983). Selection of bacteria with favorable transport properties through porous rock for the application of microbial enhanced oil recovery. *Appl. Environ. Microbiol.*, 46:1066-1072.

Jansson, P.E. and L. Karlberg (2001). Coupled heat and mass transfer model of soil-plant-atmosphere systems. Royal Institute of Technology, Department of Civil and Environmental Engineering. Stockholm: 325.

Jaques, D. and J. Šimůnek (2005). User Manual of the Multicomponent Variably-Saturated Flow and Transport Model HP1. Description, Verification and Examples. Version 1.0, SCK•CEN-BLG-998, Waste and Disposal, SCK•CEN, Mol, Belgium, 79 p.

Jin Y., and M. Flury (2002). Fate and transport of viruses in porous media. In: Advances in Agronomy, 77:39–84.

Jin, Y., Y. Chu, and Y. Li (2000). Virus removal and transport in saturated and unsaturated sand columns. *Journal of Contaminant Hydrology* 43(2): 111-128.

John, D.E. and J.B. Rose (2005). Review of Factors Affecting Microbial Survival in Groundwater. *Environ. Sci. Technol.* 39(19): 7345-7356.

Jones, C.A., CV. Cole, A. N. Sharpley, and J.R. Williams (1984). A simplified soil and plant phosphorus model: I. Documentation. *Soil Science Society of America Journal* 48: 800-805.

Jones, C.A. and J.R. Kiniry, Ed.s (1986). CERE-N Maize: a simulation model of maize growth and development. Texas A&M University Press. Temple, TX. 194p.

Jones, O.A.H., N. Voulvoulis, and J.N. Lester (2002). Aquatic environmental assessment of the top 25 English prescription pharmaceuticals. *Water Research* 36(20): 5013-5022.

Jones, O.A.H., N. Voulvoulis, and J.N. Lester (2006). Partitioning Behavior of Five Pharmaceutical Compounds to Activated Sludge and River Sediment. *Environmental Contamination and Toxicology* 50(3): 297-305.

Jonkers, N., R. Laane, C. de Graaf, and P. de Voogt (2005). Fate modeling of nonylphenol ethoxylates and their metabolites in the Dutch Scheldt and Rhine estuaries: validation with new field data. *Estuarine, Coastal, and Shelf Science* 62(1-2): 141-160.

Kadlec, R.H. and R.I. Knight (1996). Treatment Wetlands. CRC Press, Boca Raton, FL.

Kanda, R., P. Griffin, H.A. James, and J. Fothergill (2003). Pharmaceutical and personal care products in sewage treatment works. *Journal of Environmental Monitoring* 5(5): 823-830.

Karathanasis, A.D., T.G. Mueller, B. Boone, and Y.L. Thompson (2006). Effect of soil depth and texture on fecal bacteria removal from septic effluents. *Journal of Water and Health* 4(3): 395-404.

Keller, V. (2005). Risk assessment of "down-the-drain" chemicals: Search for a suitable model. *Science of the Total Environment* 360(1-3): 305-318.

Kinney, C.A., E.T. Furlong, S.L. Werner, and J.D. Cahill (2006). Presence and distribution of wastewater-derived pharmaceuticals in soil irrigated with reclaimed water. *Environmental Toxicology and Chemistry* 25(2): 317-326.

Kinzelbach, W., W. Schafer, and J. Herzer (1991). Numerical modeling of natural and enhanced denitrification processes in aquifers. *Water Resources Research* 27: 1123-1135.

Kirkland, S.L. (2001). Coupling site-scale fate and transport with watershed-scale modeling to assess the cumulative effects of nutrients from decentralized onsite wastewater systems. *Geological Engineering*. Golden, CO, Colorado School of Mines. M.S.Thesis.

Knisel, W.G. (1993). Groundwater Loading Effects of Agricultural Management Systems, Version 2.10. Dept. Publication No. 5, Biological & Agricultural Engineering Department, University of Georgia-Coastal Plain Experiment Station, Tifton: 260.

Kolpin, D.W., E.T. Furlong, M.T. Meyer, E.M. Thurman, S.D. Zuagg, L.B. Barber, and H.T. Buxton (2001). Pharmaceuticals, Hormones, and Other Organic Wastewater Contaminants in U.S. Streams, 1999-2000: A National Reconnaissance. *Environmental Science and Technology* 36(6): 1201-1211.

Kowal, N.E. (1982). *Health Effects of Land Treatment: Microbiological*. Cincinnati, OH, U.S. Environmental Protection Agency.

Kristiansen, R. (1981). Sand filter trenches for purification of septic tank effluent: II. The fate of N. *J. Environmental Quality* 10(3): 358-360.

Kroes, J.G. and J. Roelsma (1997). ANIMO 3.5 User's guide for the ANIMO version 3.5 nutrient leaching model. *DLO-Staring Centrum, Wageningen, The Netherlands* Technical Document 46.

Lance, J.C. and C.P. Gerba (1977). "Nitrogen, phosphate, and virus removal from sewage water during land filtration". *Prog. Water Technol.* 9(1): 157-166.

Lance, J.C. and C.P. Gerba (1984). Virus movement in soil during saturated and unsaturated soil. *Appl Environ Microbiol* 47(2): 335-337.

Langergraber, G. (2003). Simulation of subsurface flow constructed wetlands - results and further research needs. *Water Science and Technology* 48: 157-166.

Langergraber, G. (2004). The role of plant uptake on removal of organic matter and nutrients in subsurface flow constructed wetlands: A simulation study. p. 491-498. *In* . Lienard and H. Burnett (ed.) Proceedings of the 9th IWA Specialized Group Conference for Wetland Systems for Water Pollution Control. Avignon, France. Vol. 2:26-30.

Langergraber, G. and J. Šimůnek (2005). Modeling variably saturated water flow and multicomponent reactive transport in constructed wetlands. *Vadose Zone Journal* 4(924-938).

Larsen, M. C. (2008). Statement of Dr. Robert M. Hirsch, Associate Director For Water, U.S. Geological Survey. U.S. Department of the Interior Before the Committee on Transportation and Infrastructure Subcommittee on Water Resources and Environment. September 18, 2008.

Larson, T. A., J. Lienert, A. Joss, and H. Siegrist (2004). How to avoid pharmaceuticals in the aquatic environment. *Journal of Biotechnology* 113(1-3): 295-304.

Latch, D.E., J.L. Packer, B.L. Stender, J. VanOverbeke, W.A. Arnold, and K. McNeill (2005). Aqueous photochemistry of triclosan: Formation of 2,4-dichlorophenol, 2,8-dichlorodibenzo-p-dioxin, and oligomerization products. *Environmental Toxicology and Chemistry* 24(3): 517-525.

Lee, L.S., T.J. Strock, A.K. Sarmah, and P.S.C. Rao (2003). Sorption and Dissipation of Testosterone, Estrogens, and Their Primary Transformation Products in Soils and Sediment. *Environmental Science and Technology* 37(18): 4098-4105.

Lemonds and McCray (2003). Modeling Phosphorus Transport in the Blue River Watershed, Summit County, Colorado. *Proc. of the First Interagency Conference on Research in Watersheds, USDA, USGS, BLM, Natural Resour. Conservation Serv., U.S. EPA, U.S. Forest Serv., and Consortium of Universities for the Advancement of Hydrologic Science.*

Li, C., S. Frolking, and T.A. Frolking (1992). A model of nitrous oxide evolution from soil driven by rainfall events: 1. Model structure and sensitivity. *Journal of Geophysical Research* 97: 9759-9776.

Lienert, J., K. Gudel, and B. I. Escher (2007). Screening Method for Ecotoxicological Hazard Assessment of 42 Pharmaceuticals Considering Human Metabolism and Excretory Routes. *Environmental Science and Technology* 41(12): 4471-4478.

Lietz, A.C. and M.T. Meyer (2006). Evaluation of Emerging Contaminants of Concern at the South District Wastewater Treatment Plant Based on Seasonal Sampling Events, Miami-Dade County, Florida, 2004. *USGS Scientific Investigations Report* 2006–5240.

Lindbo, D.L., J. Lynn, K. Neal, G. Young, and A. Amoozegar (2007). Applying soil morphology to long term acceptance rate determination. *In* K. Mancl (ed.) *Eleventh individual and small community sewage systems conference proceedings*, Warwick, RI, 20-24 October, 2007, ASABE Publication 701P1107, St Joseph, MI.

Lindstroem, A., I.J. Buerge, T. Poiger, P.A. Bergqvist, M.D. Mueller, and H.R. Buser (2002). Occurrence and Environmental Behavior of the Bactericide Triclosan and Its Methyl Derivative in Surface Waters and in Wastewater. *Environmental Science and Technology* 36(11): 2322-2329.

Ling, G. and A.I. El-Kadi (1998). A lumped parameter model for N transformation in the unsaturated zone. *Water Resources Research* 34: 203-212.

Litke, D.W. (1999). Review of phosphorus control measures in the United States and their effects on water quality. W. R. I. R.-. U.S. Geological Survey, National Water Quality Assessment Program.

Lodwig, E.M. A.H. Hosie, A. Bourdes, K. Findlay, D. Allaway, R. Karunakaran, J.A. Downie, and P.S. Poole (2003). Amino-acid cycling drives nitrogen fixation in the legume-Rhizobium symbiosis. *Nature* 422: 722-726.

Logan, B. E., D. G. Jewett, R. G. Arnold, E. J. Bouwer, and C. R. O'Melia (1995). Clarification of clean-bed filtration models, *J. Environ. Eng.*, *121*, 869-873.

Lombardo, P. (2006). Phosphorus geochemistry in septic tanks, soil absorption systems, and groundwater. Newton, MA, Lombardo Associates, Inc.

Lotse, E.G., J.D. Jabro, K.E. Simmons, and D.E. Baker (1992). Simulation of N dynamics and leaching from arable soils. *Journal of Contaminant Hydrology* 10: 183-196.

Lowe, K.S., N.K. Rothe, J.M.B. Tomaras, K. DeJong, M.B. Tucholke, J. Drewes, J.E. McCray, and J. Munakata-Marr (2007). WERF project 04-DEC-1, Influent constituent characteristics of the modern waste stream from single sources: literature review. Alexandria, VA, Water Environment Research Foundation.

Lowrance, R. (1992). Groundwater nitrate and denitrification in a Coastal Plain riparian forest. *Journal of Environmental Quality* 21: 401-405.

Lowrance, R., L.S. Alter, J.D. Newbold, R.R. Schnabel, P.M. Groffman, J.M. Denver, D.L. Correll, J.W. Gillam, J.L. Robinson, R.B. Brinsfield, K.W. Staver, and A.H. Todd (1995). Water quality functions of riparian forest buffer systems in the Chesapeake Bay Watershed. *U.S. EPA* Washington, D.C.

Mackay, D., S. Paterson and W. Y. Shiu (1992). Generic models for evaluating the regional fate of chemicals. *Chemosphere* 24(6): 695-717.

Magdoff, F.R., J. Bouma, and D.R. Keeney (1974). Columns representing mound-type disposal systems for septic tank effluent: II. Nutrient transformations and bacterial populations. *J. Environmental Quality* 3: 228-234.

Mansell, J. and J.E. Drewes (2004). Fate of Steroidal Hormones During Soil-Aquifer Treatment. *Ground Water Monitoring and Remediation* 24(2).

Mansell, R.S., S.A. Bloom, B.A. Burgoa, P. Nkedi-Kizza, and J.S. Chen (1991). Experimental and simulated P transport in soil using a multireaction model. *Journal of Soil Science* 153: 185-194.

Martin, J.P. and D.D. Focht (1977). Soils for managinment of organic wastes and waste waters. *In*: L.F. Elliott and F.J. Stevenson (Editors). Soil Science Society of America, American Society of Agronomy, Crop Science Society of America, Madison, WI, pp. 650.

Matthies, M., J. Berlekamp, F. Koormann, and J.O. Wagner (2001). Georeferenced regional simulation and aquatic exposure assessment. *Water Science and Technology* 43(7): 231-238.

Maurer, M., D. Abramovich, R.L. Siegrist, and W. Gujer (1999). Kinetics of biologically induced phosphorus precipitation in waste-water treatment. *Water Research* 33(2): 484-493.

Mazzone, H.M. (1998). *CRC Handbook of Viruses,* CRC Press, Boca Raton, FL.

McAvoy, D.C., A.J. DeCarvalho, A.M. Nielsen, and M.L. Cano (2002). Investigation of an onsite wastewater treatment system in sandy soil: modeling the fate of surfactants. *Environmental Toxicology and Chemistry* 21(12): 2623-2630.

McCray, J.E., S.L. Kirkland, R.L. Siegrist, and J.D. Thyne (2005). Model parameters for simulating fate and transport of on-site wastewater nutrients. *Ground Water* 43(4): 628–639.

McLeese, D.W., V. Zitko, D.B. Sergeant, L. Burridge, and C.D. Metcalfe (1981). Lethality and Accumulation of Alkylphenols in Aquatic Fauna. *Chemosphere* 10(7): 723-730.

McLeod, M.,J. Aislabie, J. Smith, R. Fraser, A. Roberts, and M. Taylor (2001). Viral and chemical tracer movement through contrasting soils. *J Environ Qual* 30(6): 2134-40.

Mendoza, M. (2008). Water task force misses target: Still no plan to deal with drug contamination. *Seattle Post-Intelligencer*. Seattle. http://seattlepi.nwsource.com/national/358924_water14.htmlv.

Mengis, M.S.L. Schiff, M. Harris, M.C. English, R. Aravena, R.J. Elgood, and A. MacLean (1999). Multiple geochemical and isotopic approaches for assessing ground water NO_3^- elimination in a riparian zone. *Ground Water* 37(3): 448-457.

Metcalf & Eddy, Inc. (2003). *Wastewater Engineering: Treatment and Reuse-4th Edition.* Revised by G. Tchobanoglous, F. L. Burton, H. D. Stensel. McGraw-Hill Companies, Inc., New York, NY.

Metherell, A.K., L.A. Harding, C.V. Cole, and W.J. Parton (1993). CENTURY Manual: CENTURY soil organic matter model environment. *Great Plains System Research Unit* USDA-ARS, Fort Collins, CO. Technical Report No. 4.

Miettinen, I.T., T. Vartiainen, and P.J. Martikainen (1997). Phosphorus and bacterial growth in drinking water. *Applied and Environmental Microbiology* 63(8): 3242–3245.

Miles, R.J., R. Rubin, and L.T. West (2007). *Fecal coliform distribution under pressure dosed onsite wastewater systems*. Eleventh Individual and Small Community Sewage Systems Conference Proceedings Warwick, RI, ASABE.

Misra, C., D.R. Nielsen, and J.W. Biggar (1974). Nitrogen transformations in soil during leaching: I. Theoretical considerations. *Soil Science Society of America Journal* 38: 289-294.

Moldovan, Z. (2006). Occurrences of pharmaceutical and personal care products as micropollutants in rivers from Romania. *Water Research* 64(11): 1808-1817.

Molz, F.J., M.S. Widdowson, and L.D. Benefield (1986). Simulation of microbial growth dynamic coupled to nutrient and oxygen transport in porous media. *Water Resour. Res.*, 22:8, 1207-1216.

Moore, R.S., D.H. Taylor, M.M.M. Reddy, and L.S. Sturman (1982). Adsorption of reovirus by minerals and soils. *Appl Environ Microbiol* 44(4): 852-859.

Mulligan, C.N. and R.N. Yong (2004). Natural Attenuation of Contaminated Soils, *Environ. Intern.* 30, 587-601.

Myrold, D.D. (1988). Denitrification in rye grass and winter wheat cropping systems of western Oregon. *Soil Sci. Soc. Am. J.* 52:412-416.

Nair, V.D., K.M. Portier, D.A. Graetz and M.L. Walker (2004). An environmental threshold for degree of phosphorus saturation in sandy soils. *Journal of Environmental Quality* 33: 107-113.

Neitsch, S.L., J.G. Arnold, J.R. Kiniery, R. Srinivasan, and J.R. Williams (2002). Soil and Water Assessment Tool user's manual. Version 2000. Texas Water Resources Institute. College Station, TX.

Nelson, N.O. and J.E. Parsons (2007). Basic approaches to modeling phosphorus leaching. *In* D.E. Radcliffe and M.L. Cabrera (Ed.s): *Modeling phosphorus in the environment*. CRC Press. Boca Raton, FL.: pp 81-103.

Newton, B.J. and W.M. Jarrell (1999). A procedure to estimate the response of aquatic systems to changes in phosphorus and N inputs. USDA National Water and Climate Center, Natural Resources Conservation Service.

NJOSP (1988). Development of nitrate dilution model for land use planning in the state of New Jersey. *New Jersey Office of State Planning* Technical Reference Document 32.

Oaks, J.L., M. Gilbert, M.Z. Virani, R.T. Watson, C.U. Meteyer, B.A. Rideout, H.L. Shivaprasad, S. Ahmed, M.J.I. Chaudhry, and M. Arshad (2002). Diclofenac residues as the cause of vulture population decline in Pakistan. *Nature* 40: 65-81.

Orvos, D.R., D.J. Versteeg, J. Inauen, M. Capdevielle, A. Rothenstein, and V. Cunningham (2002). Aquatic Toxicity of Triclosan. *Environmental Toxicology and Chemistry* 21(7): 1338-1349.

Painter, H.A. and J.E. Loveless (1981). Effect of temperature and pH value on the growth-rate constants of nitrifying bacteria in the activated-sludge process. *Water Res.* 17(3): 237-248.

Pang, L., M. McLeod, J. Aislabie, J. Šimůnek, M. Close, and R. Hector (2008). Modeling Transport of Microbes in Ten Undisturbed Soils under Effluent Irrigation. *Vadose Zone J* 7(1): 97-111.

Park, N.S., T.N. Blandford, and P.S. Huyakorn (1992). VIRALT, Version 2.1, Documentation and User's Guide. , HydroGeoLogic, Herndon, VA.

Park, N.S., T.N. Blandford and P.S. Huyakorn (1993). CANVAS: a composite analytical-numerical model for viral and solute transport simulation. *Documentation and User's Guide, Version 1.0*, HydroGeoLogic, Herndon, VA.

Parkhurst, D.L. and C.A.J. Appelo (1999). User's guide to PHREEQC (Version 2) - A computer program for speciation, batch-reaction, one-dimensional transport, and inverse geochemical calculations. Water-Resources Investigations Report 99-4259. *Available online at ftp://brrcrftp.cr.usgs.gov/geochem/unix/phreeqc/manual.pdf* verified 20 May 2008.

Parkin, T.B. and J.A. Robinson (1989). Stochastic models of soil denitrification. *Applied and Environmental Microbiology* 55: 72-77.

Parkin, T.B., J.L. Starr, and J.M. Meisinger (1987). Influence of sample size on measurement of soil denitrification. *Soil Sci. Soc. Am. J.* 51:1492-1501.

Parry, S., P. Renault, J. Chadeuf, C. Chenu, and R. Lensi (2000). Particulate organic matter as a source of variation in denitrification in clods of soil. *European Journal of Soil Science* 51(2): 271-281.

Parton, W.J., A.R. Mosier, D.S. Ojima, D.W. Valentine, D.S. Schimel, K. Weier, and A.E. Kulmala (1996). Generalized model for N_2 and N_2O production from nitrification and denitrification. *Global Biogeochem. Cycles* 10: 401-412.

Parton, W.J., A.E. Holland, S.J.D. Grosso, M.D. Hartman, R.E. Martin, A.R. Mosier, D.S. Ojima, and D.S. Schimel (2001). General model for NO_x and N_2O emissions from soils. *J. Geophys. Res.* 106: 17,403-17,419.

Patenaude, E.L., J.A. Atoyan, D.A. Potts, and J.A. Amador (2008). Effects of tetracycline on water quality, soil and gases in aerated and unaerated leachfield mesocosms. *Journal of Environmental Science and Health Part A* in press.

Pavan, M., and A.P. Worth (2008). Review of estimation models for biodegradation. *QSAR & Combinat. Sci.*, 27:1, 32-40.

Pelish, J., D. Slusher, and L. Johnson (2003). Spectrum Laboratories Chemical Fact Sheet: Compound 106467 http://www.speclab.com/index.shtml. Ft. Lauderdale, FL & Savannah, GA.

Pennington, J.C., D. Gunnison, M. Zakikhani, H.L. Fredickson, and D.B. Ringelberg (1999). Natural attenuation of explosives in soil and water systems at department of defense sites: Final report. SERDP-99-1 (July 1999). 53p.

Pérez Estrada, L.A., S. Malato, W. Gernjak, A. Agüera, E.M. Thurman, I. Ferrer, and A.R. Fernández Alba (2005). Photo Fenton degradation of diclofenac: Identification of main intermediates and degradation pathway. *Environmental Science and Technology* 39(21): 8300-8306.

Pieper, A.P., J.N. Ryan, R.W. Harvey, G.L. Amy, T.H. Illangasekare, and D.W. Metge (1997). Transport and Recovery of Bacteriophage PRD1 in a Sand and Gravel Aquifer: Effect of Sewage-Derived Organic Matter. *Environ. Sci. Technol.* 31(4): 1163-1170.

Pinay, G., L. Roques, and A. Fabre (1996). Spatial and temporal patterns of denitrification in a riparian forest. *Journal of Applied Ecology* 30: 581-591.

Poole, S.K. and C.F. Poole (1999). Chromatographic models for the sorption of neutral organic compounds by soil from water and air. *Journal of Chromatography* 845(1-2): 381-400.

Potts, D.A., J.H. Gorres, E.L. Nicosia, and J.A. Amador (2004). Effects of aeration on water quality from septic system leachfields. *J Environ Qual* 33(5): 1828-38.

Powelson, D.K. and C.P. Gerba (1994). Virus removal from sewage effluents during saturated and unsaturated flow through soil columns. *Water Research* 28: 2175-2181.

Powelson, D.K., J.R. Simpson, and C.P. Gerba (1990). Virus Transport and Survival in Saturated and Unsaturated Flow through Soil Columns. *J Environ Qual* 19(3): 396-401.

Powelson, D.K., J.R. Simpson, and C.P. Gerba (1991). Effects of organic matter on virus transport in unsaturated flow. *Appl Environ Microbiol* 57(8): 2192-6.

Radcliffe, D.E. and M.L. Cabrera (2007). Modeling phosphorus in the environment. *CRC Press* Boca Raton, FL.

Radcliffe, D.E. and L.T. West (2009). Design hydraulic loading rates for on-site wastewater systems. *Vadose Zone Journal* 8: 64-74.

Rajagopalan, R., and C. Tien (1976). Trajectory analysis of deep-bed filtration with the sphere in-cell porous media model, *AIChE. J.*, 22: 523-533.

Rand, G.M. (1995). *Fundamentals of Aquatic Toxicology: Effects, Environmental Fate, and Risk Assessment*, Taylor & Francis.

Rassam, D., D. Pagendam, and H. Hunter (2005). The riparian nitrogen model (RNM) - Basic theory and conceptualization. Cooperative Research Centre for Catchment Hydrology, 05/9.

Rezania, L. (2004). Optimizing phosphate treatment to minimize lead/copper seasonal variations. *Waterline*, Minnesota Department of Health, Environmental Health Division. 12.

Rice, C.W. and M.S. Smith (1982). Denitrification in no-till and plowed soils. *Soil Sci. Soc. Am. J.* 46:1168-1173.

Richardson, V.L. and W.S. Janna (2006). Performance between three types of subsurface wastewater systems using Hydrus-2D. 13 p. Available at www.soil.ncsu.edu/lockers/lindbo/Nowra%202006%20Proceedings/Papers/PE-06-39.pdf

Roberts, P., J.P. Roberts, and D.L. Jones (2006). Behaviour of the endocrine disrupting chemical nonylphenol in soil: Assessing the risk associated with spreading contaminated waste to land. *Soil Biology and Biochemistry* 38(7): 1812-1822.

Robertson, W.D. (1994). Chemical fate and transport in a domestic septic system: site description and attenuation of dichlorobenzene. *Environmental Toxicology and Chemistry* 13(2): 183-191.

Robertson, W.D. (1995). Development of steady-state phosphate concentrations in septic system plumes. *Journal of Contaminant Hydrology* 19: 289-305.

Robertson, W. D. (2003). Enhanced attenuation of septic system phosphate in noncalcarious sediments. *Ground Water* 41:48-56.

Robertson, W.D. and J. Harman (1999). Phosphate plume persistence at two decommissioned septic system sites. *Ground Water* 37: 228-236.

Robertson, W.D., S.L. Schiff, and C.J. Ptacek (1998). Review of phosphate mobility and persistence in 10 septic system plumes. *Ground Water* 33(6): 1000-1010.

Ronner, A.B. and A.C.L. Wong (1994). *Microbial clogging of wastewater infiltration systems.* Seventh International Symposium on Individual and Small Community Septage Systems, Atlanta, GA, ASAE.

Rothe, N. (2006). Comparison of raw wastewater and septic tank effluent composition through literature and field investigations. *ESE*. Golden, CO, Colorado School of Mines. MS.

Ruskin, R. (1992). Reclaimed water and subsurface irrigation. ASAE Paper. No 92-2578. St. Joseph Michigan.

Ryan J.N. and M. Elimelech (1996). Colloid mobilization and transport in groundwater. *Colloids Surfaces A: Physicochem Eng Aspects* 107, pp. 1–56.

Ryden, J.C. (1983). Denitrifcation loss from a grassland soil in the field receiving different rates of N as ammonium nitrate *Journal of Soil Science* 34: 355-365.

Ryther, J.H. and W.M. Dunstan (1971). Nitrogen, phosphorus and eutrophication in the coastal marine environment. *Science* 171: 1008-1013.

Saadoun, I., Z. Al-Ghazawi, Q. Ababneh, J. Schijven, H. van den Berg, J. Al-Rashdan, M. Shawaqfeh, W. Blandford, T. Boving, and M. Meqdam (2008). Removal of fecal indicator coliform and bacteriophages by riverbank filtration (RBF) in Jordan. Proceedings of the International Conference on Construction and Building Technology (ICCCBT), Kuala Lumpur, Malaysia.

Sabaliunas, D., S.F. Webb, A. Hauk, M. Jacob, and W.S. Eckhoff (2003). Environmental fate of Triclosan in the River Aire Basin, UK. *Water Research* 37(13): 3145-3154.

Saggar, S., Bolan, N.S., Bhandral, R., Hedley, C.B. and Luo, J. (2004). A review of emissions of methane, ammonia, and nitrous oxide from animal excreta deposition and farm effluent application in grazed pastures. *New Zealand Journal of Agricultural Research*, 47: 513-544.

Samsøe-Petersen, L., M. Winther-Nielsen, and T. Madsen (2003). Fate and effects of triclosan. *Environmental Project No. 861*. D. W. a. Environment, Danish Environmental Protection Agency.

Saxton, K.E. and W.J. Rawls (2004). Soil water characteristics by texture and organic matter for hydrologic solutions. *Soil Science Society of America annual conference*. Seattle, WA.

Scandura, J.E. and M.D. Sobsey (1997). Viral and bacterial contamination of groundwater from on-site sewage treatment systems. *Water Science and Technology* 35(11-12): 141-146.

Schaap, M.G., F.J. Leij, and M.T. van Genuchten (2001). Rosetta: a computer program for estimating soil hydraulic parameters with hierarchical pedotransfer functions. *J. of Hydrology* 251: 163-176.

Schaub, S.A. and C.A. Sorber (1977). Virus and Bacteria Removal from Wastewater by Rapid Infiltration Through Soil. *Appl Environ Microbiol* 33(3): 609-619.

Scheytt, T., P. Mersmann, M. Leidig, A. Pekdeger, and T. Heberer (2004). Transport of Pharmaceutically Active Compounds in Saturated Laboratory Columns. *Ground Water* 42(5): 767-773.

Scheytt, T.J., P. Mersmann, and T. Heberer (2006). Mobility of pharmaceuticals carbamazepine, diclofenac, ibuprofen, and propyphenazone in miscible-displacement experiments. *Journal of Contaminant Hydrology* 83(1-2): 53-69.

Schijven, J.F., and S.M. Hassanizadeh (2000). Removal of viruses by soil passage: overview of modeling, processes, and parameters. *Critical Reviews of Environ. Sci. Technol.,* 30(1):49-127.

Schindler, D.W. (1977). Evolution of phosphorus limitation in lakes. *Science* 195: 260–262.

Schwarzenbach, R.P. and J. Westall (1981). Transport of nonpolar organic compounds from surface water to groundwater. Laboratory sorption studies. *Environmental Science and Technology* 15(11): 1360-1367.

Sears, C.L. (2005). A dynamic partnership: Celebrating our gut flora. *Anaerobe* 11(5): 247-251.

Seitz Jr, E.P., A.L. Waggoner, P.S. Fox, and T.J. Taylor (2005). Compositions having enhanced deposition of a topically active compound on a surface, Google Patents.

Sekiguchi, Y., Matsunaga, A., Yamamoto, A. and Inoue, Y. (2000). Analysis of condensed phosphates in food products by ion chromatography with an on-line hydroxide eluent generator. *Journal of Chromatography* A, 881: 639-644.

Selim, H.M. and I.K. Iskandar (1981). Modeling N transport and transformations in soils: 2. Validation. *Soil Science* 131: 303-312.

Shaffer, M.J. (1995). Fate and transport of N: What models can and cannot do. *NRCS* Working Paper No. 11.

Shaffer, M.J., A.D. Halyorson, and F.J. Pierce (1991). Nitrate leaching and economic analysis package (NLEAP): Model description and application. Proceedings of a Symposium sponsored by Division A-5 of the American Society of Agronomy: 285-322.

Sharpley, A.N. (2000). Bioavailable phosphorus in soil. *In* G.M. Pierzynski (ed.), Methods for Phosphorus Analysis for Soils, Sediments, Residuals, and Waters. Southern Cooperative Series Bulletin No. 396, p. 39-45.

Sharpley, A.N., C.A. Jones, C. Gray, and C.V. Cole (1984). A simplified soil and plant phosphorus model: II. Prediction of labile, organic, and sorbed phosphorus. *Soil Science Society of America Journal* 48(805-809).

Shuval, H.I. and N. Gruener (1972). Epidemiological and toxicological aspects of nitrates and nitrites in the environment. *American Journal of Public Health* 62: 1045-1051.

Siegrist, R.L. (1987). Soil clogging during subsurface wastewater infiltration as affected by effluent composition and loading rate. *J. Environmental Quality* 16(2): 181-187.

Siegrist, R.L. (2007). Engineering of a soil treatment unit as a unit operation in an onsite wastewater system, *In* K. Mancl (ed.) *Eleventh individual and small community sewage systems conference proceedings*, Warwick, RI, 20-24 October, 2007, ASABE Publication 701P1107, St Joseph, MI.

Siegrist, R.L. and W.C. Boyle (1987). Wastewater Induced Soil Clogging Development. *Journal of Environmental Engineering*. 113(3): 550-566.

Siegrist, R.L., E.J. Tyler, and P.D. Jenssen (2001). Design and Performance of Onsite Wastewater Soil Absorption Systems. National Research Needs Conference: Risk-Based

Decision Making for Onsite Wastewater Treatment. EPRI report no. 1001446, Electric Power Research Institute, Palo Alto, CA.

Siegrist, R.L., K. Lowe, J.E. McCray, B. Beach, and S. Van Cuyk (2002). *Accelerated loading for evaluation of long-term performance of soil PMBs used for wastewater renovation.* 11th Northwest On-Site Wastewater Short Course and Exhibition. April 3-4, 2002, Seattle, WA.

Silverbush, J. and S.A. Barber (1983). Prediction of phosphorus and potassium uptake by soybeams with a mechanistic mathematical model. *Soil Science Society of America Journal* 47: 262-265.

Sim, Y., and C.V. Chrysikopoulos (2000). Virus transport in unsaturated porous media. *Water Resour. Res.* 36(1):173-179.

Šimůnek, J., M. Sejna, and M.T. van Genuchten (1991). The HYDRUS-2D software package for simulating two-dimensional movement of water, heat, and multiple solutes in variably saturated media. Version 2.0. 53: 251.

Šimůnek, J., K. Huang, M. Šejna, and M. van Genuchten (1998). The HYDRUS-1D software package for simulating one-dimensional movement of water, heat, and multiple solutes in variably-saturated media. Version 2.0. IGWMC-TPS-70. International Ground Water Modeling Ctr., Colorado School of Mines, Golden, CO.

Šimůnek, J., M.T. Van Genuchten, and M. Senja (2005). The HYDRUS-1D software package for simulating the one-dimensional movement of water, heat, and multiple solutes in variably-saturated media. University of California, Riverside, CA, Department of Environmental Science.

Šimůnek, J., M.T. Van Genuchten and M. Senja (2006). The HYDRUS software package for simulating two- and three-dimensional movement of water, heat, and multiple solutes in variably-saturated media. Version 1.0. Technical Manual. PC Progress, Prague.

Šimůnek, J., M. Šejna, and M. van Genuchten (2007). The HYDRUS Software Package for Simulating Two- and Three-Dimensional Movement of Water, Heat, and Multiple Solutes in Variably-Saturated Media, User Manual, Version 1.0, PC Progress, Prague, Czech Republic, 2007.

Singh, L. (2007). An experiment to identify levels of triclosan that are harmful to bacteria in Puget Sound. *Triclosan effects on marine bacteria*, University of Washington.

Skaggs, R.W. (1978). A water management model for shallow water table soils. Report 134. Raleigh, NC: North Carolina State University, North Carolina Water Resources Research Institute.

Sobsey, M.D., C.H. Dean, M.E. Knuckles, and R.A. Wagner (1980). Interactions and survival of enteric viruses in soil materials. *Appl Environ Microbiol* 40(1): 92-101.

Sobsey, M.D. and P.A. Shields (1987). Survival and transport of viruses in soils: model studies. *Human viruses in sediments, sludges, and soils*. V. C. Rao, Melnick, J.L. Boca Raton, FL, CRC Press: 155-177.

Spiteri, C., C.P. Slomp, P. Regnier, C. Meilie, and P. Van Cappellen (2007). Modeling the geochemical fate and transport of wastewater-derived phosphorus in contrasting groundwater systems. *Journal of Contaminant Hydrology* 92: 87-108.

Sposito, G. (1989). *The Chemistry of Soils*. New York, Oxford University Press. book

Stevenson, F.J. (1994). *Humus Chemistry Genesis, Composition, Reactions*. New York, NY, John Wiley & Sons.

Stevik, T.K., G. Ausland, J.F. Hanssen, and P.D. Jenssen (1999a). The influence of physical and chemical factors on the transport of E. coli through biological filters for wastewater purification. *Water Research* 33(18): 3701-3706.

Stevik, T.K., G. Ausland, P.D. Jenssen, and R.L. Siegrist (1999b). Removal of E. coli during intermittent filtration of wastewater effluent as affected by dosing rate and media type. *Water Research* 33(9): 2088-2098.

Stevik, T.K., A. Kari, G. Ausland, and J. Fredrik Hanssen (2004). Retention and removal of pathogenic bacteria in wastewater percolating through porous media: a review. *Water Research* 38(6): 1355-1367.

Strauss, E.A., N.L. Mitchell, and G.A. Lamberti (2002). Factors regulating nitrification in aquatic sediments: effect of organic carbon, N availability, and pH. *Canadian Journal of Fisheries and Aquatic Science* 59: 554-563.

Strenn, B., M. Clara, O. Gans, and N. Kreuzinger (2004). Carbamazepine, diclofenac, ibuprofen and bezafibrate--investigations on the behaviour of selected pharmaceuticals during wastewater treatment. *Water Science and Technology* 50(5): 269-76.

Suarez, S., M. Dodd, F. Omil, and U. von Gunten (2007). Kinetics of triclosan oxidation by aqueous ozone and consequent loss of antibacterial activity: Relevance to municipal wastewater ozonation. *Water Research* 41: 2481-2490.

Suller, M.T.E. and A.D. Russell (2000). Triclosan and antibiotic resistance in Staphylococcus aureus. *Antimicrobial Agents and Chemotherapy* 46(1): 11-18.

Sundaram, K.M. and S. Szeto (1981). The dissipation of nonylphenol in stream and pond water under simulated field conditions. *Environmental Science and Health* 16(6): 767-76.

Taylor, J.R. (2003). Evaluating groundwater nitrates from on-lot septic systems, and guidance model for land planning in Pennsylvania. Pennsylvania State University. State College. PA.

Thompson, A., P. Griffin, R. Stuetz, and E. Cartmell (2005). The Fate and Removal of Triclosan during Wastewater Treatment. *Water Environment Research* 77(1): 63-67.

Thompson, J.R., H.R. Sorenson, H. Gavin, and A. Refsgaard (2004). Application of the couples MIKE SHE/MIKE 11 modelling system to a lowland wet grassland in southeast England. *Journal of Hydrology* 293(1): 151-179.

Thompson, S.S. and M.V. Yates (1998). Bacteriophage inactivation at the air–water–solid interface in dynamic batch systems. *Appl Environ Microbiol* 65(3): 1186-1190.

Tiedje, J.M. (1988). Ecology of denitrification and dissimilatory nitrate reduction to ammonium. *In*: A.J.B. Zehnder (Ed.), *Biology of Anaerobic Microorganisms.* pp179-244.

Tixier, C., H.P. Singer, S. Oellers, and S.R. Mueller (2003). Occurrence and Fate of Carbamazepine, Clofibric Acid, Diclofenac, Ibuprofen, Ketoprofen, and Naproxen in Surface Waters. *Environmental Science and Technology* 37(6): 1061-1068.

Topp, E. and A. Starratt (2000). Rapid mineralization of the endocrine-disrupting chemical 4-nonylphenol in soil. *Environmental Toxicology and Chemistry* 19(2): 313-318.

Toride, N., F.J. Leij, and M.T. Van Genuchten (1995). The CXTFIT code for estimating transport parameters from laboratory or field tracer experiments, version 2.1. *Research Report from the U.S. Salinity Laboratory* 137.

Trouwborst, T., S. Kuyper, J.C. De Jong, and A.D. Plantinga (1974). Inactivation of some bacterial and animal viruses by exposure to liquid-air interfaces. *Journal of General Virology* 24: 155-165.

Tucholke, M. (2007). Statistical assessment of relationships between denitrification and easily measurable soil properties: a simple predictive tool for watershed-scale modeling. *Environmental Science and Engineering*. Golden, CO, Colorado School of Mines. M.S. Thesis.

Tufenkji, N. and M. Elimelech (2004). Correlation Equation for Predicting Single-Collector Efficiency in Physicochemical Filtration in Saturated Porous Media, *Environ. Sci. Technol.*, 38, 529-536.

Ullman, J.L. (2007). The chemical behavior of estrone and 17beta-estradiol in the environment. *Biological & Agricultural Engineering*, Texas A&M University. Doctorate.

U.S. Census Bureau (2005). Supplemental measures of material well-being: Basic needs, consumer durables, energy, and poverty, 1981 to 2002. E. a. S. A. U.S. Department of Commerce.

U.S. Environmental Protection Agency (1997). Response to Congress on Use of Decentralized Wastewater Treatment Systems. U.S. Environmental Protection Agency, Office of Water, Washington, D.C.

U.S. Environmental Protection Agency (1999). The report to Congress: Summaries of state ground water conditions. A groundwater report to Congress. Office of Water Supply and Office of Solid Waste Management Programs, Washington, D.C.

U.S. Environmental Protection Agency (2000). National Primary Drinking Water Regulations: Ground Water Rule; Proposed Rule. U.S. Environmental Protection Agency, Office of Water, Washington, D.C

U.S. Environmental Protection Agency (2001). Protocol for developing pathogen TMDLs. EPA 841-R-00-002, Office of Water (4503F), United States Environmental Protection Agency, Washington, D.C., 2001. p. 132.

U.S. Environmental Protection Agency (2002). *Onsite Wastewater Treatment Systems Manual*. Cincinnati, OH, U.S. Environmental Protection Agency.

U.S. Environmental Protection Agency. (2008a). Drinking Water Contaminants (February 15, 2008). Retrieved April 7, 2008, http://www.epa.gov/safewater/contaminants/index.html.

U.S. Environmental Protection Agency (2008b). Land Treatment of Municipal Wastewater Effluents: Process Design Manual. EPA/625/R06/016. National Risk Management Research Laboratory, Cincinnati, OH.

U.S. Food and Drug Administration. (1992). The Bad Bug Book - Foodborne Pathogenic Microorganisms and Natural Toxins Handbook. Retrieved May 5, 2008, 2008.

Valiela, I., J.L. Bowen, and K.D. Kroeger (2002). Assessment of models for estimation of land-derived N loads to shallow estuaries. *Applied Geochemistry* 17: 935-953.

Van Cuyk, S. and R.L. Siegrist (2001). Pathogen fate in wastewater soil absorption systems as affected by effluent quality and soil clogging genesis. *On-Site Wastewater Treatment*. Fort Worth, TX, ASAE.

Van Cuyk, S. and R.L. Siegrist (2007). Virus removal within a soil infiltration zone as affected by effluent composition, application rate, and soil type. *Water Res* 41(3): 699-709.

Van Cuyk, S., R. Siegrist, A. Logan, S. Masson, E. Fischer, and L. Figueroa (2001). Hydraulic and purification behaviors and their interactions during wastewater treatment in soil infiltration systems. *Water Research* 35(4): 953-964.

Van Cuyk, S., R.L. Siegrist, K. Lowe, and R.W. Harvey (2004). Evaluating microbial purification during soil treatment of wastewater with multicomponent tracer and surrogate tests. *Journal of Environmental Quality* 33: 316-329.

van Genuchten, M.T. (1980). A closed-form equation for predicting the hydraulic conductivity of unsaturated soils. *Soil Science Society of America Journal*, 44: 892-898.

van Genuchten, M. T. and J. C. Parker (1984). Boundary conditions for displacement experiemnts through short laboratory soil columns. *Soil Science Society of America Journal* 48: 703-708.

Vaughn, J.M., E.F. Landry, C.A. Beckwith, and M.Z. Thomas (1981). Virus removal during groundwater recharge: effects of infiltration rate on adsorption of poliovirus to soil. *Appl Environ Microbiol* 41(1): 139-147.

Vaughn, J.M., E.F. Landry, and M.Z. Thomas (1983). Entrainment of Viruses from Septic Tank Leach Fields Through a Shallow, Sandy Soil Aquifer. *Appl Environ Microbiol* 45(5): 1474-1480.

Ver Hey, M.E. and W.W. Woessner (1987). *Documentation of the degree of waste treatment provided by septic systems, vadose zone and aquifer in intermontane soils underlain by sand and gravel*. Proceedings of the Fifth National Symposium on individual and small community sewage treatment, Chicago, IL.

Verstraeten, I.M., G.S. Fetterman, M.T. Meyer, T. Bullen, and S.K. Sebree (2005). Use of tracers and isotopes to evaluate vulnerability of water in domestic wells to septic waste. *Ground Water Monitoring and Remediation* 25(2): 107-117.

Vinyard, D.L. and M.H. Bates (1979). High-calcium flyash for tertiary phosphorus removal. *Water and Sewage Works* 6: 62-104.

Walker, W.G., J. Bouma, D.R. Keeney, and F.R. Magdoff (1973). Nitrogen transformations during subsurface disposal of septic tank effluent in sands: II. Groundwater Quality. *J. Environmental Quality* 2: 521-525.

Wang, X., M.A. Youssef, R.W. Skaggs, J.D. Atwood, D.H. Franklin, and J.R. Frankenberger. (2005). Sensitivity analysis of the nitrogen simulation model DRAINMOD-N II. *Trans of ASAE* 48:2205-2212.

Ward, J.H.J. (1963). Hierarchical grouping to optimize an objective function. *Journal of the American Statistical Association* 58(301): 236-244.

Ward, M.H., S.D. Mark, K.P. Cantor, D.D. Weisenburger, A. Correa-Villasenor, and S.H. Zahm (1996). Drinking water nitrate and the risk of Non-Hodgkin's Lymphoma. *Epidemiology* 7(5): 465-471.

Weintraub, L.H., R.A. Goldstein, and R.L. Siegrist (2004). Watershed modeling of onsite wastewater systems in Summit County, Colorado. *Water Environment Federation.*

Weiskel, P.K. and B.L. Howes (1992). Differential transport of sewage-derived N and phosphorus through a coastal watershed. *Environmental Science and Technology* 26(2): 352-360.

Whelan, B.R. (1988). Disposal of septic tank effluent in calcareous sands. *J. Environmental Quality* 17(2): 272-277.

Williams, J.R. (1995). The EPIC model. *In* V.J. Singh (ed.) Computer Models of Watershed Hydrology. Water Resources Publications, Highlands Ranch, CO: 909-1000.

Williams, J.R., P.T. Dyke, and C.A. Jones (1983). EPIC: a model for assessing the effects of erosion on soil productivity. *In* W.K. Lauenroth (Ed.), *Analysis of Ecological Systems: State-of-the-Art in Ecological Modeling*, Elsevier, Amsterdam. pp. 553-572.

Wilson, J.T., C.G. Enfield, W.J. Dunlap, R.L. Cosby, D.A. Foster, and L.B. Baskin (1981). Transport and fate of selected organic pollutants in a sandy soil. *Journal of Environmental Quality* 10(4).

Wong-Chong, G.M. and R.C. Loehr (1975). The kinetics of microbial nitrification. *Water Res.* 9(12): 1099-1106.

World Health Organization (W.H.O.) (2006). Guidelines for drinking-water quality [electronic resource] : incorporating first addendum. Vol. 1, Recommendations. – 3rd ed.

Yalkowsky, S. H. and Y. He (2003). Handbook of Aqueous Solubility Data. *CRC Press.*

Yates, M.V. and C.P. Gerba (1998). Microbial considerations in wastewater reclamation and reuse. *Wastewater Reclamation and Reuse*. T. Asano. Lancaster, PA, Technomic. 10**:** 437-488.

Yao, K.M., M.T. Habibian, and C.R. O'Melia (1971). Water and waste water filtration: concepts and applications, *Environ. Sci. Technol.* 5, 1105-1112.

Yates, M.V. and W.A. Jury (1995). On the use of virus transport modeling for determining compliance. *J Environ Qual.* 24:1051–1055.

Yates, M.V. and Y. Ouyang (1992). VIRTUS, a model of virus transport in unsaturated soils. Appl. and Environ. Microbio. 55:5, 1609-1616.

Yeh, G-T., J-R. Cheng, and T.E. Short (1997). 3DFATMIC: User's Manual of a Three-Dimensional Subsurface Flow, Fate and Transport of Microbes and Chemicals Model - Version 1.0. EPA/600/R-97/053, Washington DC.

Ying, G.G. and R.S. Kookana (2007). Triclosan in wastewaters and biosolids from Australian wastewater treatment plants. *Environmental International* 33(2): 199-205.

Ying, G.G., X.Y. Yu, and R.S. Kookana (2007). Biological degradation of triclocarban and triclosan in a soil under aerobic and anaerobic conditions and comparison with environmental fate modelling. *Environmental Pollution* 150: 300-305.

Youssef, M.A., R.W. Skaggs, G.M. Chescheir, and J.W. Gilliam (2005). The nitrogen simulation model, DRAINMOD-N II. *Trans ASAE* 48:611-626.

Yu, Z., B. Xiao, W. Huang, and P. Peng (2004). Sorption of Steroid Estrogens to Soils and Sediments. *Environmental Toxicology and Chemistry* 23(3): 531-539.

Zimmerman, M.J. (2004). Occurrence of Organic Wastewater Contaminants, Pharmaceuticals, and Personal Care Products in Selected Water Supplies, Cape Cod, Massachusetts. *U.S. Geological Survey Scientific Open-File Report 2005* 1206: 16.

Zurawsky, M.A., W. D. Robertson, C.J. Ptacek, and S.L. Schiff (2004). Geochemical stability of phosphorus solids below septic system infiltration beds. *Journal of Contaminant Hydrology* 73: 129-143.

Zwiener, C. and F.H. Frimmel (2003). Short-term tests with a pilot sewage plant and biofilm reactors for the biological degradation of the pharmaceutical compounds clofibric acid, ibuprofen, and diclofenac. *Science of the Total Environment* 309(1-3): 201-211.

WERF SUBSCRIBERS

WASTEWATER UTILITY

Alabama
Montgomery Water Works & Sanitary Sewer Board

Alaska
Anchorage Water & Wastewater Utility

Arizona
Avondale, City of
Glendale, City of, Utilities Department
Mesa, City of
Peoria, City of
Phoenix Water Services Dept.
Pima County Wastewater Management
Safford, City of
Tempe, City of

Arkansas
Little Rock Wastewater Utility

California
Central Contra Costa Sanitary District
Corona, City of
Crestline Sanitation District
Delta Diablo Sanitation District
Dublin San Ramon Services District
East Bay Dischargers Authority
East Bay Municipal Utility District
El Dorado Irrigation District
Fairfield-Suisun Sewer District
Fresno Department of Public Utilities
Inland Empire Utilities Agency
Irvine Ranch Water District
Las Gallinas Valley Sanitary District
Las Virgenes Municipal Water District
Livermore, City of
Los Angeles, City of
Los Angeles County, Sanitation Districts of
Napa Sanitation District
Novato Sanitary District
Orange County Sanitation District
Palo Alto, City of
Riverside, City of
Sacramento Regional County Sanitation District
San Diego Metropolitan Wastewater Department, City of
San Francisco, City & County of
San Jose, City of
Santa Barbara, City of
Santa Cruz, City of
Santa Rosa, City of
South Bayside System Authority
South Coast Water District
South Orange County Wastewater Authority
Stege Sanitary District
Sunnyvale, City of
Union Sanitary District
West Valley Sanitation District

Colorado
Aurora, City of
Boulder, City of
Greeley, City of
Littleton/Englewood Water Pollution Control Plant
Metro Wastewater Reclamation District, Denver

Connecticut
Greater New Haven WPCA
Stamford, City of

District of Columbia
District of Columbia Water & Sewer Authority

Florida
Broward, County of
Fort Lauderdale, City of
Jacksonville Electric Authority (JEA)
Miami-Dade Water & Sewer Authority
Orange County Utilities Department
Pinellas, County of
Reedy Creek Improvement District
Seminole County Environmental Services
St. Petersburg, City of
Tallahassee, City of
Tampa, City of
Toho Water Authority
West Palm Beach, City of

Georgia
Atlanta Department of Watershed Management
Augusta, City of
Clayton County Water Authority
Cobb County Water System
Columbus Water Works
Fulton County
Gwinnett County Department of Public Utilities
Savannah, City of

Hawaii
Honolulu, City & County of

Idaho
Boise, City of

Illinois
Decatur, Sanitary District of
Greater Peoria Sanitary District
Kankakee River Metropolitan Agency
Metropolitan Water Reclamation District of Greater Chicago
Wheaton Sanitary District

Iowa
Ames, City of
Cedar Rapids Wastewater Facility
Des Moines, City of
Iowa City

Kansas
Johnson County Wastewater
Lenexa, City of
Unified Government of Wyandotte County/ Kansas City, City of

Kentucky
Louisville & Jefferson County Metropolitan Sewer District

Louisiana
Sewerage & Water Board of New Orleans

Maine
Bangor, City of
Portland Water District

Maryland
Anne Arundel County Bureau of Utility Operations
Howard County Bureau of Utilities
Washington Suburban Sanitary Commission

Massachusetts
Boston Water & Sewer Commission
Massachusetts Water Resources Authority (MWRA)
Upper Blackstone Water Pollution Abatement District

Michigan
Ann Arbor, City of
Detroit, City of
Holland Board of Public Works
Saginaw, City of
Wayne County Department of Environment
Wyoming, City of

Minnesota
Rochester, City of
Western Lake Superior Sanitary District

Missouri
Independence, City of
Kansas City Missouri Water Services Department
Little Blue Valley Sewer District
Metropolitan St. Louis Sewer District

Nebraska
Lincoln Public Works and Utilities Department

Nevada
Henderson, City of
Las Vegas, City of
Reno, City of

New Jersey
Bergen County Utilities Authority
Ocean County Utilities Authority
Passaic Valley Sewerage Commissioners

New York
New York City Department of Environmental Protection

North Carolina
Charlotte/Mecklenburg Utilities
Durham, City of
Metropolitan Sewerage District of Buncombe County
Orange Water & Sewer Authority
University of North Carolina, Chapel Hill

Ohio
Akron, City of
Butler County Department of Environmental Services
Columbus, City of
Metropolitan Sewer District of Greater Cincinnati
Montgomery, County of
Northeast Ohio Regional Sewer District
Summit, County of

Oklahoma
Oklahoma City Water & Wastewater Utility Department
Tulsa, City of

Oregon
Albany, City of
Clean Water Services
Eugene, City of
Gresham, City of
Portland, City of Bureau of Environmental Services
Water Environment Services

Pennsylvania
Hemlock Municipal Sewer Cooperative (HMSC)
Philadelphia, City of
University Area Joint Authority

South Carolina
Charleston Water System
Mount Pleasant Waterworks & Sewer Commission
Spartanburg Water

Tennessee
Cleveland Utilities
Murfreesboro Water & Sewer Department
Nashville Metro Water Services

Texas
Austin, City of
Dallas Water Utilities
Denton, City of
El Paso Water Utilities

WERF SUBSCRIBERS

Fort Worth, City of
Houston, City of
San Antonio Water System
Trinity River Authority

Utah
Salt Lake City Corporation

Virginia
Alexandria Sanitation Authority
Arlington, County of
Fairfax County
Hampton Roads Sanitation District
Hanover, County of
Henrico, County of
Hopewell Regional Wastewater Treatment Facility
Loudoun Water
Prince William County Service Authority
Richmond, City of
Rivanna Water & Sewer Authority

Washington
Everett, City of
King County Department of Natural Resources
Seattle Public Utilities
Sunnyside, Port of
Yakima, City of

Wisconsin
Green Bay Metro Sewerage District
Kenosha Water Utility
Madison Metropolitan Sewerage District
Milwaukee Metropolitan Sewerage District
Racine, City of
Sheboygan Regional Wastewater Treatment
Wausau Water Works

Australia
ACTEW (Ecowise)
South Australian Water Corporation
South East Water Limited
Sydney Water Corporation
Water Corporation of Western Australia

Canada
Edmonton, City of/Edmonton Waste Management Centre of Excellence
Lethbridge, City of
Regina, City of, Saskatchewan
Toronto, City of, Ontario
Winnipeg, City of, Manitoba

New Zealand
Watercare Services Limited

United Kingdom
Anglian Water Services, Ltd.

STORMWATER UTILITY

California
Fresno Metropolitan Flood Control District
Los Angeles, City of, Department of Public Works
Monterey, City of
San Francisco, City & County of
Santa Rosa, City of
Sunnyvale, City of

Colorado
Aurora, City of
Boulder, City of

Florida
Orlando, City of

Iowa
Cedar Rapids Wastewater Facility
Des Moines, City of

Kansas
Overland Park, City of

Kentucky
Louisville & Jefferson County Metropolitan Sewer District

Maine
Portland Water District

North Carolina
Charlotte, City of, Stormwater Services

Pennsylvania
Philadelphia, City of

Tennessee
Chattanooga Stormwater Management

Texas
Harris County Flood Control District, Texas

Washington
Bellevue Utilities Department
Seattle Public Utilities

STATE

Connecticut Department of Environmental Protection
Kansas Department of Health & Environment
New England Interstate Water Pollution Control Commission (NEIWPCC)
Ohio River Valley Sanitation Commission
Urban Drainage & Flood Control District, CO

CORPORATE

ADS Environmental Services
Advanced Data Mining International
AECOM
Alan Plummer & Associates
Alpine Technology Inc.
Aqua-Aerobic Systems Inc.
Aquateam–Norwegian Water Technology Centre A/S
ARCADIS
Associated Engineering
Bernardin Lochmueller & Associates
Black & Veatch
Blue Water Technologies, Inc.
Brown & Caldwell
Burgess & Niple, Ltd.
Burns & McDonnell
CABE Associates Inc.
The Cadmus Group
Camp Dresser & McKee Inc.
Carollo Engineers Inc.
Carpenter Environmental Associates Inc.
CET Engineering Services
CH2M HILL
CRA Infrastructure & Engineering
CONTECH Stormwater Solutions
D&B/Guarino Engineers, LLC
Damon S. Williams Associates, LLC
Ecovation
EMA Inc.
Environmental Operating Solutions, Inc.
Environ/The ADVENT Group, Inc.
Fay, Spofford, & Thorndike Inc.
Freese & Nichols, Inc.
ftn Associates Inc.
Gannett Fleming Inc.
Garden & Associates, Ltd.
Geosyntec Consultants
GHD
Greeley and Hansen LLC
Hazen & Sawyer, P.C.
HDR Engineering Inc.
HNTB Corporation
Hydromantis Inc.
HydroQual Inc.
Infilco Degremont Inc.
Jacques Whitford NAWE, Inc.
Jason Consultants LLC Inc.
Jordan, Jones, & Goulding Inc.
KCI Technologies Inc.
Kelly & Weaver, P.C.
Kennedy/Jenks Consultants
KMK Consultants
Larry Walker Associates
Limno-Tech Inc.
The Low Impact Development Center Inc.
Malcolm Pirnie Inc.
Material Matters
McKim & Creed
MPR Engineering Corporation, Inc.
MWH
NTL Alaska, Inc.
O'Brien & Gere Engineers Inc.
Odor & Corrosion Technology Consultants Inc.
Paradigm Environmental Technologies, Inc.
Parametrix Inc.
Parsons
Post, Buckley, Schuh & Jernigan
Praxair, Inc.
Ring Industrial Group
RMC Water & Environment
Ross & Associates Ltd.
SAIC
Siemens Water Technologies
The Soap & Detergent Association
Smith & Loveless, Inc.
Southeast Environmental Engineering, LLC
Stearns & Wheler, LLC
Stone Environmental Inc.
Stratus Consulting Inc.
Synagro Technologies Inc.
Tetra Tech Inc.
Trojan Technologies Inc.
Trussell Technologies, Inc.
Uni-Bell PVC Pipe Association
URS Corporation
Westin Engineering Inc.
Wright Water Engineers
Zoeller Pump Company

INDUSTRY

American Electric Power
American Water
Anglian Water Services, Ltd.
Chevron Energy Technology
The Coca-Cola Company
Dow Chemical Company
DuPont Company
Eastman Chemical Company
Eli Lilly & Company
Johnson & Johnson
Merck & Company Inc.
Procter & Gamble Company
Suez Environment
United Utilities North West (UUNW)
United Water Services LLC

Note: List as of 2/21/09

IWA Publishing's authorised EU representative for General Product Safety Regulations is Diane D'Arras, 15 rue Duret, 75116 Paris, France, e-mail: safety@iwap.co.uk.

Printed and bound by CPI Group (UK) Ltd, Croydon, CR0 4YY
12/05/2026
02108881-0003